Simply Complexity

Neil Johnson is the head of a new inter-disciplinary research group in Complexity at the University of Miami in Florida. He was previously Professor of Physics and co-director of research collaboration into Complexity at Oxford University. He does research on a wide variety of real-world Complex Systems, and is an author of *Financial Market Complexity* (Oxford University Press, 2003). He enjoys the complex things in life, like overcrowded bars and the fastest route home from work.

Simply Complexity

A Clear Guide to Complexity Theory

Neil F. Johnson

ONEWORLD

A Oneworld Book

First published in hardcover by Oneworld Publications
as *Two's Company, Three is Complexity* 2007
First published in trade paperback as *Simply Complexity* 2009
Reprinted 2009, 2012 (twice)

A CIP record for this title is available
from the British Library

ISBN 978–1–85168–630–8
ebook ISBN 978–1–78074–049–2

Typeset by Jayvee, Trivandrum, India
Cover design by D. R. ink
Printed and bound in the US by RR Donnelley

Oneworld Publications
10 Bloomsbury Street
London WC1B 3SR
England
www.oneworld-publications.com

Contents

Preface

It is 2050, and you are watching *Who Wants to be a Billionaire?* The contestant is one question away from the jackpot. Up comes his question: "What is the name of the theory that scientists started developing at the beginning of the twenty-first century, and which helped the world overcome traffic congestion, financial market crashes, terrorist attacks, pandemic viruses, and cancer?" The contestant cannot believe his luck. What an easy question! But he is so nervous that his mind temporarily goes blank. He starts to consider option A: "They are all still unsolved problems" – but then quickly realizes that this is a dumb answer. Instead, he uses his last lifeline to ask the audience. The audience responds unanimously and instantaneously with option B: "The Theory of Complexity". Without hesitation, he goes with option B. The host hands him the cheque, and the world has yet another billionaire.

Pure fantasy? Maybe not.

In this book, we will go on a journey to the heart of Complexity, an emerging science which looks set to trigger the next great wave of advances in everything from medicine and biology through to economics and sociology. Complexity Science also comes with the prospect of solving a wide range of important problems which face us as individuals and as a Society. Consequently, it is set to permeate through every aspect of our lives.

There is, however, one problem. We don't yet have a fully-fledged "theory" of Complexity. Instead, I will use this book to assemble all the likely ingredients of such a theory within a common framework, and then analyze a wide range of real-world applications within this same common framework. It will then require someone from the future – perhaps one of the younger readers of this book – to finally put all these pieces into place.

Complexity Science is a double-edged sword in the best possible sense. It is truly "big science" in that it embodies some of the hardest, most fundamental and most challenging open problems in academia. Yet it also manages to encapsulate the major practical issues which face us every day from our personal lives and health, through to global security. Making a pizza is complicated, but not complex. The same holds for filling out your tax return, or mending a bicycle puncture. Just follow the instructions step by step, and you will eventually be able to go from start to finish without too much trouble. But imagine trying to do all three at the same time. Worse still, suppose that the sequence of steps that you follow in one task actually depends on how things are progressing with the other two. Difficult? Well, you now have an indication of what Complexity is all about. With that in mind, now substitute those three interconnected tasks for a situation in which three interconnected people each try to follow their own instincts and strategies while reacting to the actions of the others. This then gives an idea of just how Complexity might arise all around us in our daily lives.

While I was writing this book, I had the following "wish-list" in my head concerning its goals:

1. To provide a book which a wide cross-section of people would want to read and would enjoy reading – regardless of age, background or level of scientific knowledge.
2. To introduce readers to the exciting range of real-world scenarios in which Complexity Science can prove its worth.
3. To provide the book on Complexity that "I never had but always needed". In other words, to provide an easily readable yet thorough guide to this important scientific revolution.
4. To provide a book that my kids could read – or rather, a book that they would actually *choose* to read all by themselves. This is a very important goal, since Complexity will likely become *the* science of interest for future generations.

5. To provide a book which is just as readable on a plane or bus as in a library. As such, it should also make sense when read in short chunks.
6. To provide a book which provides professional scientists, economists, and policy-makers with a new perspective on open problems in their field, and to help stimulate new Complexity-based interdisciplinary research projects.

However, as I finish the book and offer it up to potential readers, I realize that the above wish-list can essentially be reduced to just one item: I would wish that you enjoy reading this book, and that it might provide you with fresh thoughts and insights for dealing with the complex world in which we live, and which our children will inherit.

There are some practicalities concerning the book's content and layout which I would like to explain. The language, examples and analogies are kept simple since the focus of the book is to explain *what* Complexity Science is all about, and *why* it is so important for us all. I therefore avoid delving into too much detail in the main text. Instead, the Appendix describes how to access the technical research papers upon which the discussions in the book are based, and gives a list of Internet websites containing additional information about Complexity research around the world. Having said this, I won't pull any punches in the sense that I tackle all the topics which I believe to be relevant. Part 1 of the book takes us through the theoretical underpinnings of Complexity, while Part 2 delves into its real-world applications. Some of the territory is only just beginning to be explored, with very few answers available for the questions being posed. From the perspective of other scientific revolutions throughout history this might seem to be par for the course. However we are not talking about history here – instead, we are looking at work which is emerging at the forefront of a new discipline. For this reason we will be highlighting where such research is heading, rather than where it has been.

But why should you believe what I write about Complexity? This is a crucially important question given that Complexity Science is still being developed and its potential applications explored. Unfortunately many accounts of Complexity in the popular press are second-hand, i.e. they are typically written by people who have done little, if any, research on Complexity themselves and are instead reporting on their interpretation of other people's work. Given the relatively immature nature of the field, I believe that such indirect interpretations are potentially dangerous. For this reason, I will base the book's content around my own research group's experience in Complexity. This has various advantages: (i) it reflects my own understanding of the Complexity field; (ii) it represents what I believe to be the most relevant and important topics; (iii) it will hopefully give the reader a sense of what it is like to be at the "pit-face" in such a challenging area of research; and (iv) it ensures that any reader can challenge me directly on any claims that I make, and can demand an informed answer. To facilitate this process of public scrutiny, a complete list of the relevant scientific research reports is presented in the latter part of the Appendix. I also encourage any readers who wish to email me with questions, to do so at n.johnson@physics.ox.ac.uk

Finally I would like to thank most warmly the following highly talented scientists with whom I am fortunate to enjoy ongoing interactions on Complexity: Pak Ming Hui, Luis Quiroga, Ferney Rodriguez, Mike Spagat, Jorge Restrepo, Elvira Maria Restrepo, Roberto Zarama, Derek Abbott, Chiu Fan Lee, Tim Jarrett, Alexandra Olaya Castro, David Smith, Sean Gourley, Sehyo Charley Choe, Douglas Ashton, Mark McDonald, Omer Suleman, Nachi Gupta, Nick Jones, Ben Burnett, Alex Dixon, Tom Cox, Juan Pablo Calderon, Juan Camilo Bohorquez, Dan Reinstein, Mark Rondeau, Paul Summers, Stacy Williams, Dan Fenn, Richard Ecob, Adrian Flitney, Matt Berryman, Mark Fricker, Philip Maini, Sam Howison, Tim Halpin-Healy, David Wolpert and Kagan Tumer. In particular, I would like to specifically mention Felix Reed-Tsochas and Janet Efsthatiou, who are also my fellow co-directors in Oxford University's inter-departmental complex systems research group. Many of the above-named scientists have played a fundamental role in the research discussed in this book – I have indicated their contributions explicitly where appropriate. I am also very grateful to Marsha Filion at Oneworld Publications, for her constructive comments on how to finalize the manuscript – and to my mother and father for gently encouraging me to get a move on and finally finish it.

I would like to express my deepest gratitude to Elvira Maria, Daniela, Nicholas and Dylan. Thank you for putting up with a very complex husband/father while this book was being prepared, and thank you for putting last Christmas on hold.

Oxford, U.K.
2007

PART 1

What exactly *is* Complexity Science?

Chapter 1

Two's Company, Three is Complexity

1.1 A definition, of sorts

Take a look in many dictionaries, and you will find Complexity defined along the lines of "The behavior shown by a Complex System". Then look up "Complex System", and you will probably see "A system whose behavior exhibits Complexity". So what's going on? Well, unfortunately, Complexity is not easy to define. Worse still, it can mean different things to different people. Even among scientists, there is no unique definition of Complexity. Instead, the scientific notion of Complexity – and hence of a Complex System – has traditionally been conveyed using particular examples of real-world systems which scientists believe to be complex.

This book will take the "complex" out of Complexity, by going to the heart of what connects together all real-world Complex Systems. We will uncover the magic ingredients which make something complex as opposed to just being complicated, and show how Complexity is deeply engrained in our own everyday lives. We will also see why Complexity is set to revolutionize our understanding of science, and help resolve some of the most challenging problems facing society as a whole.

Complexity can be summed up by the phrase "Two's company, three is a crowd". In other words, Complexity Science can be seen as the *study of the phenomena which emerge from a collection of interacting objects* – and a crowd is a perfect example of such an *emergent phenomenon,* since it is a phenomenon which emerges from a collection of interacting people. We only have to look at world history to realize that it is riddled with major events which have been driven by human crowd behavior. Everyday examples of crowds include collections of commuters, financial market traders, human cells, or insurgents – and the associated crowd-like phenomena which emerge are traffic jams, market crashes, cancer tumors, and guerilla wars. Even extreme weather conditions such as floods, heatwaves, hurricanes, and droughts can be seen as a sort of crowd effect, since they emerge from the collective behavior of "packets" of water and air in the form of oceans, clouds, winds and air moisture. And if we add to this the collective actions of humans – in particular, the environmental changes caused by human activity – we conjure up the controversial emergent phenomenon known as "global warming".

1.2 Complexity in action

At the heart of most real-world examples of Complexity, is the situation in which a collection of objects are competing for some kind of limited resource – for example, food, space, energy, power, or wealth. In such situations, the emergence of a crowd can have very important practical consequences. For example, in a financial market, or the housing market, the spontaneous formation of a crowd of people who wish to sell – and hence are effectively competing for buyers – can lead to a market crash in which the price falls dramatically in a short time. A related crowd phenomenon occurs among commuters who are competing for space on a particular road at the same time. This leads to a traffic jam, which is the traffic equivalent of a market crash. Other examples include Internet overloads and power blackouts, in which subscribers simultaneously decide to access and hence exhaust the available resources of a particular computer system or power network. Even wars and terrorism can be viewed as the collective, violent actions of different groups of people who are fighting for control of the same resources, e.g. land or political power.

The Holy Grail of Complexity Science is to understand, predict and control such emergent phenomena – in particular, potentially catastrophic crowd-like effects such as market crashes, traffic jams, epidemics, illnesses such as cancer, human conflicts, and environmental change. Are they predictable in any way, or do they just appear out of nowhere without warning? Can they be controlled, manipulated or even avoided?

What is remarkable about such emergent phenomena, is that they can arise in the absence of any central controller or coordinator. Just think about the level of coordination and communication which some central controller would actually require in order to be able to recreate a particular traffic jam. In other words, imagine the number of cell-phone calls he would have to make to ensure that all the drivers were on the same road at the same time, and in one particular pattern. It simply couldn't be done in a reliable way. This represents a universal feature of Complex Systems: emergent phenomena can arise without the need for an "invisible hand". Instead, the collection of objects is able to self-organize itself in such a way that the phenomenon appears all by itself – as if by magic.

The sheer power and momentum of these emergent phenomena can also be quite remarkable. We all know how easy it is to be swept up in the ebbs and flows of mob mentality – whether intentionally or unintentionally. Recent decades such as the 1970s delivered cultural tsunamis in terms of fashions and hairstyles: just think flared trousers and platform shoes. In the 1990s, we had the infamous dot-com boom with company employees agreeing to be paid in stock options rather than hard cash – only to find themselves penniless when the bubble burst around April 2000. And who hasn't had the experience of wandering along a busy street in the middle of a crowd of people, only to find yourself separated from your companions and going in a direction you don't actually want to go? We each seem to have an innate urge to join in with a crowd – but it may not be the best decision from our individual perspective. Just think of selling or buying a house or car. You will get a far better price if you sell when everybody else is buying, and vice versa.

It is not just collections of people that show emergent phenomena. The animal, insect and fish kingdoms are awash with examples of self-organization: from ant-trails and wasp swarms through to bird flocks and fish schools. In fact, biology is sitting on a treasure-chest of such collective phenomena – from the immune system's collective response to invading viruses through to intercellular communication and signalling which drives many important biological processes. The fact that all these effects represent emergent phenomena explains why so many different disciplines are getting interested in Complexity.

Closer to everyone's personal concerns – and indeed, worries – is the area of human health and medicine. This is a prime example of Complexity in action. Our immune system consists of a collection of defense mechanisms for dealing with invading viruses. However just like the traffic, the stock market and the Internet, the system can go wrong all by itself – for example, when the collective response of the immune system ends up attacking healthy tissue. Hence understanding the extent to which we can predict, manage and even control a Complex System has particular importance from the perspective of human health. Indeed it may even lead to new forms of treatment whereby the collective responses of the body are harnessed to deal with a specific problem in a particular organ, rather than relying on one particular targeted therapy. A cancer tumor is a particularly horrific example of a crowd effect gone wrong. Instead of staying in check, cells begin to multiply uncontrollably – and just as with other Complex System phenomena such as traffic jams, it becomes very hard to know what to do to reduce the size of the tumor without causing some even more damaging secondary effects. For example, any treatment which involves damaging the tumor may indirectly lead to the survival of the fittest, most malignant cells.

Interest in Complexity is not confined to natural objects such as people, animals or cells. The ability of a collection of objects to produce emergent phenomena without the need for some central controller, has attracted the attention of researchers at NASA. In particular, Kagan Tumer and David Wolpert have been leading a research team at Ames Research Laboratory in Mountain View, California which is looking at emergent phenomena in collections of machines. The machines in question could be robots, satellites, or even micro-spacecraft. For example, NASA are investigating the possibility that a collection of relatively simple robots can be used to explore the surface of a planet in a fast and efficient manner – as opposed to using one large and far more complicated machine. They have a good reason for doing this. If one robot in this collection were to malfunction, there would still be plenty more available. By contrast, a single malfunction in the large machine could lead to the immediate termination of a very costly mission. This also explains NASA's interest in exploring the properties of collections of simple satellites, as opposed to one large sophisticated one – and also collections of micro-spacecraft as opposed to one much larger one.

But there is another, far more intriguing reason that NASA is interested in such research. Most NASA missions are likely to involve sending machines to distant planets – and it is hard to maintain reliable communication channels over such distances. It would therefore be wonderful if NASA engineers could just sit back, relax and let the machines on the planet sort it out for themselves. This would of course land the machines with the same difficulty as we have when trying to arrange a lunch-date by phone with a group of friends. Judging from what typically happens with the lunch-date problem, you might think that one of the machines would simply end up acting as the local coordinator, checking one-by-one the position and availability of each machine and then coordinating their actions. This sounds like it should work fine – however, the collection of machines would then be reduced to having the same vulnerability as a single sophisticated machine. If the local coordinator malfunctions, the mission is once again over. Instead, the "killer application" aspect of such a collection of machines, and hence the interest in such Complex Systems within NASA, is that it is not necessary for the machines to have local coordination in order for them to do a good job. It turns out that a suitably chosen collection of such objects can work *better* as a group if they are not being coordinated by some single controller, but are instead competing for some limited resource – which is actually NASA's case, since there will typically be relatively few loose rocks available for picking up within a given area of a planet's surface.

A busy shopping mall provides a nice everyday example of why such a collection of selfish machines could be so useful.

Imagine that you have dropped a one-hundred dollar bill. You organize a search-team, stating that they will all share the money when it is found. If the search-team is a large one, you will have great difficulty in coordinating everybody's actions – hence you might never find the money. By contrast, if you tell everyone that the money is theirs if they find it, their individual selfish drive will likely be so strong that the money is found very quickly. In the sense that dropped bills are like available rocks, we can see that the collective action of selfish machines could be used to solve quite a complicated search problem.

There are even research groups investigating how such a collection of machines might design itself, by allowing the individual machines to adapt and evolve of their own accord. This research borrows ideas from real-world situations involving collections of humans. After all, humans acting in the setting of a financial market are doing nothing other than competing for a limited resource in a selfish way – exactly like the machines. The same applies for drivers in traffic: it is because of their competition for space on a road that we typically see arrangements of cars which are spread out in some reasonably regular pattern.

Now, if you are reading this book on a plane, you might want to take a deep breath. The increasingly high-tech nature of on-board computer systems means that each next-generation aircraft will itself be a Complex System – a Complex System which needs to be managed and controlled. But as well as creating a challenge in itself, ideas from Complexity are being harnessed to develop novel designs for such aircraft. For example, Ilan Kroo and co-workers at Stanford University have been looking at lining the back of conventional aircraft wings with a collection of robotic microflaps. The design is such that the flaps compete to be orientated in the right direction at the right time, according to the plane's planned trajectory – just like our selfish shoppers would compete to be in the right place at the right time in order to pick up the lost money. A central controller, which in this context is an aircraft pilot, would therefore no longer be needed. Now, the possibility of pilot-less planes might sound scary, but apparently many people would indeed be willing to fly in such an aircraft as long as it is cheap – and as long as their bags turn up on time.

And while we are in the air, what about those air conditions? More generally, what about the effects of our own collective actions on our environment and weather? Global competition for increasingly scarce natural resources is leading to increased levels of pollution and deforestation, and these may in turn affect our climate. The weather results from a complicated ongoing interaction between the atmosphere and oceans, connected as they are by currents of water, winds and air moisture. Floods, hurricanes, and droughts represent extreme phenomena which emerge from this collective behavior. Although scientists know the mathematics which describes individual air and water molecules, building up a picture of what billions of them will do when mixed together around the globe is extremely complicated. Now add on top of this the collective actions of human beings, and you come up against the emergent monster of global warming – and in particular, the complex question of evaluating how the Earth's climate is affected by the collective actions of its inhabitants, and what can then be done about it.

So that is Complexity in action – from technology, to health, to everyday life. But does it play any role in fundamental science, and in particular fundamental Physics? Well, it turns out that it does – and in a very big way. When you get down to the level of atoms, the range of emergent phenomena is simply breathtaking. Electrons are negatively charged particles which typically orbit the nucleus in an atom. However if you put together a large collection of such electrons, you will uncover a wealth of exotic crowd effects: from superconductivity through to effects such as the so-called Fractional Quantum Hall Effect and Quantum Phase Transitions.

It doesn't stop there. If we take just two particles such as electrons, they can show a particular type of "quantum crowd effect" called entanglement. This is an emergent phenomenon which is so bizarre that it kept Einstein baffled for the whole of his life. Indeed the information processing power underlying such a quantum crowd is so powerful that it has led to proposals for a quantum computer, which is a fundamentally new type of computer that is light years ahead of any conventional PC; quantum cryptography, which can yield completely secure secret codes; and quantum teleportation. There is even the possibility that such effects are already being exploited by Mother Nature herself – but more of that in chapter 11.

Even the fundamental physics of Einstein's space–time and Black Holes doesn't escape the hidden clutches of Complexity. At the very heart of Einstein's theories of relativity was the idea that space and time are coupled together. Another way of saying the same thing is that two pieces of space and time can interact with each other by means of light passing between the two. Hence the entire fabric of space–time is a complicated network of interconnected pieces. In chapter 5 we will look more closely at networks in general – suffice to say that they are just another way of representing a set of objects that are interacting, i.e. they are just another way of representing a Complex System.

In all of these examples, the precise nature of the crowd-like phenomena which emerge will depend on how the individual objects interact and how interconnected they are. It is extremely difficult, if not impossible, to deduce the nature of these emergent phenomena based solely on the properties of an individual object. For this reason, it is pretty much true that every new crowd effect which is found involving fundamental quantum particles such as electrons, leads to a Nobel Prize in Physics. Even though we understand the properties of a single electron, for example, the corresponding emergent phenomena from a collection of them tend to be so surprising that each one represents a remarkable new discovery by itself. On an everyday level, we know that market crashes and traffic jams can also be surprising – both in their form and in terms of when they occur and how long they last. Given this difficulty in predicting what crowd effects will arise, under what conditions and when, we can begin to see how Complexity Science might also be referred to as the science behind surprise.

So it seems like Complexity has many possible applications across the sciences, medicine and in our everyday world. Whether you are interested in fundamental physics, biology, human health, or you just want to avoid traffic jams on your way home from work, Complexity is key.

1.3 Why is my own life so complex?

It is 6 p.m. You are leaving work – and the only thing on your mind is to get home quickly. But which route should you take? It turns out you have a choice. But so does everybody else. And this is the point: the best route is the one which is the least crowded – but it is the collective decisions of everyone else which determine which of the possible routes this turns out to be. In effect you are not deciding between routes home – you are instead trying to out-guess everyone else. In other words, you are trying to out-guess the crowd in the competition for space on the road. Of course, everyone else is trying to do the same. Thinking back to our earlier discussion, this everyday situation represents an ideal candidate Complex System since it comprises a collection of objects (drivers) competing for a limited resource (road space).

But your complex life doesn't stop there. You get home, eventually, and decide you would like to go out to relax. You want to go to a particular bar – but let's assume that this bar has a limited capacity and so not everyone who turns up may actually get in. You yet again find yourself having to decide which choice to make: do you make the effort to get ready, get to the bar and run the risk that you won't get in? Or do you stay at home and run the risk that you are missing a great night out? Since the bar has a limited capacity, and yet is so popular that there are lots of potential attendees, you are again trying to out-guess the crowd. In particular, you are trying to predict whether the bar will be over-capacity or not, and hence what your action should be. Everyone else is trying to do the same. So this is again an ideal candidate Complex System since it comprises a collection of objects (bar-goers) competing for a limited resource (a place in the bar).

Say you decide not to go. Instead you will cook a nice meal at home. But you need to buy food. Where should you go? There are two supermarkets, one called "zero" and the other called "one", on opposite sides of the town. Which will be least crowded? It is again the same situation of competition for a limited resource – in this case, space in the market.

Things don't get better when, following the meal, you decide to go online and review the stock that you bought a year ago. You get the price chart up on your screen. The stock's price has gone up and down – but what is that telling you? Should you buy more stock, or sell the stock you already have? Suppose you decide to sell. If everyone else also decides to sell, there will be a sudden oversupply of these shares. Nobody would then pay you very much for them. On the other hand if you manage to sell at a moment when there are lots of buyers, you will be laughing. The same holds for selling things on any other market, from housing through to eBay. Even though you may be buying or selling based on some long-term preference or need, the decision of exactly when to buy or sell is a strategic one – and is dominated by the need to predict what everyone else will do. In other words, you must once again try to out-guess the crowd. Everybody else is again trying to do the same, and obviously not everybody can win. As a result, we once more have an ideal candidate system for Complexity since we have a collection of objects (investors) competing for a limited resource (a favorable price).

When you start to think about it, there are loads of examples from our everyday lives where, in one form or another, we are indirectly trying to out-guess what everyone else will do. And unfortunately for all of us, the correct action in such situations is determined by what everybody else actually does. What is worse is that such everyday problems are repeated over and over again, as each day goes past. This then tempts us to try to learn from the past and hence adapt our strategies to try to improve our chances of coming out on top. In other words, our daily life becomes a sequence of ongoing games – a sort of multiple "rat race".

This common everyday situation in which a collection of objects (people) repeatedly compete for some kind of limited resource, illustrates the complexities of everyday life extremely well – a fact that was first pointed out by Brian Arthur and later by John Casti, both of the Santa Fe Institute in New Mexico. But even more remarkable is the fact that it also provides us with a generic Complex System which can be adapted to describe a wide range of scientific, medical, and technological scenarios. We have already discussed various applications in section 1.2 in connection with the design of collections of machines – and as we move further through the book, we will see this same generic set-up reappearing in various guises.

1.4 The key components of Complexity

There is no rigorous definition of Complexity. But that isn't so bad – after all, it is hard to define a word such as "happiness" and yet we all know what its characteristics are. We will characterize Complexity in a similar way by describing the features which a Complex System should have, and looking at the behaviors which it should then show. This might sound very abstract – but fortunately the everyday scenarios that we have discussed come straight to our rescue. Indeed it will turn out that these characteristics are the very same ones that make our own everyday lives so complex.

Most Complexity researchers would agree that any candidate Complex System should have most or all of the following ingredients:

The system contains a collection of many interacting objects or "agents". In the case of markets, these are traders or investors. In the case of traffic, these are drivers. Typically the scientific community refers to such objects as agents. Interactions between these agents may arise because the agents are physically close to each other, or because they are members of some sort of group, or because they share some common information. For example, the agents may be linked together by some public information that they share – like investors who are watching the same price chart for a given stock, or commuters who are listening to the same traffic report on the radio. On the other hand, some agents may be linked together by private information, like two investors who happen to be friends sharing private information over the phone. To the extent that the agents are linked together through their interactions, they can also be thought of as forming part of a network. For this reason, networks have become an integral part of Complexity Science, together with the study of collections of agents. Indeed for many scientists in the community, the study of Complexity is synonymous with the study of agents and networks together.

These objects' behavior is affected by memory or "feedback". This means that something from the past affects something in the present, or that something going on at one location affects what is happening at another – in other words, a sort of knock-on effect. For example, if you happened to have taken Route 0 home for the past few nights and it was always overcrowded, you may choose to flip to Route 1 tonight. Hence you have used information from the past to influence your decision in the present – in other words, the past has been fed back into your present decision. Of course the nature of this feedback can change with time. For example, you may care less about past outcomes if it is the start of the week as opposed to the end of it. The net result of everyone having such memory can be that the system as a whole also remembers. In other words, a particular global pattern or sequence appears in the traffic or in the stock market.

The objects can adapt their strategies according to their history. This simply means that an agent can adapt its behavior by itself, in the hope of improving its performance.

The system is typically "open". This means that the system can be influenced by its environment, just like a market might be affected by outside news about the earnings of a particular company – or the traffic is affected by the closure of a particular road. By contrast, a closed system means one which is not in contact with the outside world – sort of like an office on a desert island with no Internet. And just like it sounds, such truly closed systems are rare. Much more common are systems that in some way are in contact with the outside world. In fact, the only truly closed system is the Universe as a whole. The trouble is, as we will see in chapter 2, that most fundamental theories in Physics only apply to closed systems. This is one reason why Complex Systems are so interesting not just to engineers, biologists and social scientists, but also to theoretical physicists.

The resulting system – a Complex System – will then show the following behaviors, all of which are characteristic of Complexity:

The system appears to be "alive". The system evolves in a highly non-trivial and often complicated way, driven by an ecology of agents who interact and adapt under the influence of feedback. For example, financial analysts often talk as though the market were a living, breathing object, assigning it words such as pessimistic or bearish, and confident or bullish.

The system exhibits emergent phenomena which are generally surprising, and may be extreme. In scientific terminology, the system is far from equilibrium. This basically means that anything can happen – and if you wait long enough, it generally will. For example, all markets will eventually show some kind of crash, and all traffic systems will eventually have some kind of jam. Such phenomena are generally unexpected in terms of when they arise – hence one aspect of surprise. But the system will also tend to exhibit emergent phenomena which are themselves surprising in that they could not have been predicted based on a knowledge of the properties of the individual objects. For example, no amount of understanding of the properties of water molecules could have led to the prediction that an iceberg would form and sink the *Titanic* as it passed. In terms of emergent phenomena such as market crashes and traffic jams, an important question concerns whether these extreme events might result from a sort of comedy of errors, like one domino knocking over another. For example, in the animated movie "Robots!" one small domino falling over eventually leads to a tidal wave of dominos – a sort of domino tsunami – upon which Mr. Bigwell and the other robots ended up surfing.

The emergent phenomena typically arise in the absence of any sort of "invisible hand" or central controller. In other words, a Complex System can evolve in a complicated way all by itself. For this reason, Complex Systems are often regarded as being more than the sum of their parts which is just another way of saying "Two's company, three is a crowd". Given that the Universe itself is a Complex System of sorts, this feature deals a damaging blow to proponents of so-called Intelligent Design.

The system shows a complicated mix of ordered and disordered behavior. For example, traffic jams arise at a particular point in time and at a particular place on a road network, and then later disappear. More generally, all Complex Systems seem to be able to move between order and disorder of their own accord. Put another way, they seem to exhibit pockets of order. We return to this point later in the book.

1.5 Complexity: the Science of all Sciences

But what is the value added by Complexity? After all, Complexity Science is only really of value if it can add new insights or lead to new discoveries – for example, by uncovering connections between phenomena which were previously considered unrelated. There is no point inventing a new name if we are just repackaging things that we already know. You might, for example, think that all the things that scientists traditionally look at are already sufficiently complicated to qualify as Complexity Science. As we shall see in later chapters, it is certainly true that many of the systems which scientists already study could be labelled as complex according to our list. However, the *way* in which scientists have traditionally looked at these systems does not use any of the insight of Complexity Science. In particular, the connections between such systems have not been properly explored – particularly between systems taken from different disciplines such as biology and sociology. Indeed it is fascinating to see if any insight gained from having partially understood one system, say from biology, can help us in a completely different discipline, say economics. One particular example of this is the ongoing research of Mark Fricker, Janet Efstathiou and Felix Reed-Tsochas at Oxford University, in which they analyze the nutrient supply-lines in a fungus in order to see whether lessons can be learned for supply-chain design in the retail trade.

In an everyday context, the negative effect of overlooking similarities between supposedly unrelated systems, is akin to someone becoming an expert on the detailed cultural life of New York, Washington, and Boston – yet never realizing that these cities have a shared culture because of their location on the East Coast of the United States. Unfortunately such bridge-building is doubly difficult in a scientific context, because no individual scientist can possibly know the details of all the other research fields which might be relevant. This not only holds up the advance of Complexity

Science as a whole, but it also reduces the chances of new breakthroughs in our understanding of important real-world systems.

Much of traditional Physics has dealt with trying to understand the microscopic details within what we see. This has led to physicists smashing open atoms to look at the bits inside, and then smashing these bits open to see the bits inside the bits – eventually getting down to the level of quarks. It is certainly complicated – but this reductionist approach is in a sense the opposite of what Complexity is all about. Instead of smashing things apart to find out what the components are, Complexity focuses on what new phenomena can emerge from a collection of relatively simple components. In other words, Complexity looks at the complicated and surprising things which can emerge from the interaction of a collection of objects which themselves may be rather simple. Hence the philosophical questions driving Complexity Science are similar to those for the manufacturers of a toy like LEGO: starting with a set of quite simple objects, what can I make out of them, and what complicated and surprising things can I make them do? And what happens if I change one piece for another, does that change the types of things I can make? If I am missing a few pieces, or I add a few specialist pieces, how does that change the spectrum of possible things that can be built?

Going further, the underlying philosophy behind the search for a quantitative theory of Complexity is that we don't need a full understanding of the constituent objects in order to understand what a collection of them might do. Simple bits interacting in a simple way may lead to a rich variety of realistic outcomes – and that is the essence of Complexity.

Complexity therefore represents a slap in the face for traditional reductionist approaches to understanding the world. For example, even a detailed knowledge of the specifications of a car's engine, colour and shape, is useless when trying to predict where and when traffic jams will arise in a new road system. Likewise, understanding individuals' personalities in a crowded bar would give little indication as to what large-scale brawls might develop. Within medical science, it is likely that no amount of understanding of an individual brain cell is likely to help us understand how to prevent or cure Alzheimer's disease.

So what have we got so far? We have seen why Complexity is likely to be important not only for many areas of science, but also across many other disciplines and indeed everyday life. In particular, we have seen that its role in making connections between previously unrelated phenomena taken from distinct scientific disciplines is likely to be a very important one. For this reason, we can justifiably think of Complexity as a sort of umbrella science – or even, the Science of all Sciences.

Chapter 2

Disorder rules, OK?

One of the tell-tale characteristics that a particular system, such as traffic or a financial market, is complex is that it exhibits emergent phenomena which are surprising, extreme and self-generated – just think of a traffic jam or a financial market crash. Although it is certainly true that some traffic jams and market crashes are triggered by a particular outside event (for example, a road accident or the announcement of a particular company going bankrupt), more often than not there is no obvious reason either for their appearance or disappearance. In particular, they are not being engineered or controlled by some mysterious "invisible hand" operating in the background. So what makes them appear and disappear of their own accord?

We've all had the experience of driving happily along an apparently clear highway, only to suddenly find ourselves in a traffic jam for no apparent reason. And then, just as mysteriously, the jam disappears. We drive on, looking for an obvious cause for the jam such as an accident – but there is none. The same happens in financial markets, where it is actually quite rare that the cause of a given market crash can be assigned to a particular event or set of events. Indeed, for every financial expert who says that the cause of a given crash was X, you can find one who says it was Y. For example, the dot-com bubble which burst around April 2000 was supposedly "bound to happen". But why did it happen at that specific time? And if these experts were so certain that it would happen, why were they unable to predict it beforehand?

This remarkable ability of a Complex System to generate changes in its own behavior, means that a Complex System can appear to be jogging along quite happily in a fairly random way – and then all of a sudden it exhibits extreme behavior analogous to a traffic jam or market crash. There are many real-world systems which are sufficiently complex that they also show such extreme behavior; for example, cell-phone networks where the flow of data-packets plays the role of a flow of cars, and computer systems where the demand from users plays the role of the demand from traders. Even our own bodies are sufficiently complex that such extreme changes can arise; for example, a heart attack, an epileptic fit, a seizure, and a collapsed immune system, are all examples of a sudden, spontaneous and unexpected collective action within the body. Regardless of which of these phenomena we consider most relevant to our own lives, it is clearly very important to get to the bottom of what causes such extreme behavior – and then work out if we can predict it, control it, and possibly even avoid it.

But there is something very strange going on here. Phenomena like a traffic jam and a market crash are actually quite ordered effects, since they involve a collection of otherwise independent objects suddenly locking together in some fairly synchronized way. And yet they somehow emerge out of the everyday disorder of traffic and markets for no apparent reason – like a phoenix rising from the flames. For example, the appearance of a traffic jam means that a large number of cars that had previously been dotted all over the road in typical everyday traffic style, suddenly all become lined up in one single, slow-moving mass; and a market crash means that a financial market which had previously been filled with people buying and selling in apparently random fashion, suddenly becomes filled with people who have all decided to sell at the same time. What's more, such effects can then disappear just as suddenly. So what is going on?

Complex Systems are able to move spontaneously back and forth between ordered behavior such as a traffic jam or a market crash, and the disorder typical of everyday operation, *without any external help*. In other words, a Complex System can move freely between disorder and order, and back again, and can therefore be said to exhibit "pockets of order". The emergence of such pockets of order has very important implications in terms of being able to predict and control the system. Their appearance is also quite mysterious – after all, if a bag of unsorted socks were a Complex System (which it isn't) it should therefore be capable of organizing itself into an ordered pile of pairs, ready for placing in the clothes cupboard. A wonderful idea but as we all know it doesn't happen in something as simple as a collection of socks. So there must be something more complicated going on at the heart of a Complex System which we need to understand. But the good news in practical terms is that pockets of order *can* indeed arise in a Complex System and this gives us hope that there might be a way of partially predicting the future evolution of such a system, and even being able to manage or control it.

These pockets of order can arise in both time and space. For example, traffic jams arise at a particular time and place, and then later disappear. Market crashes also arise at a particular time and in a particular world market, and then later disappear. The challenge in this chapter is to understand *why* such pockets of order arise. But to do this, we need to go on a journey from order to disorder – and where better to begin than a typical day at the office.

2.1 Another day at the office

Why is it so hard to organize our desks? Or office? Or schedule? And why is it that after a few months' use, even the most cared-for computer seems to run into all sorts of problems with file conflicts? The answer is simple: "disorder rules".

Let's look more closely at what this means. Take any set of organized objects, for example, the files in your office. Assuming you are very good at your job, these files will likely be arranged in a very specific way. Imagine that you have two files forming a pile on a shelf in your filing cabinet, and that you are

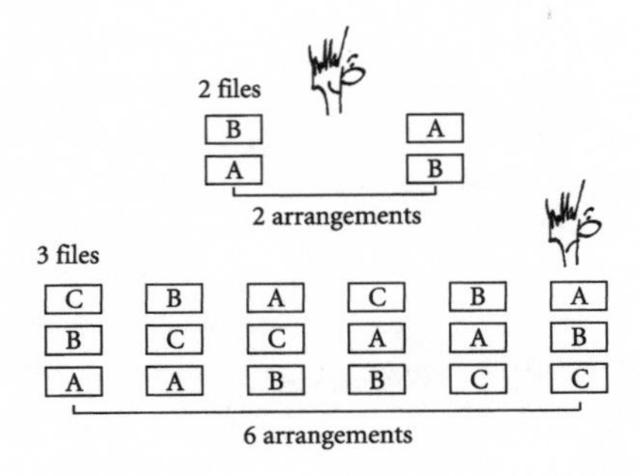

Figure 2.1 The number of possible arrangements for a pile containing (top) two files labelled A and B and (bottom) three files labelled A, B and C

unfortunate enough to have been assigned a summer intern who is careless.

Two files, one shelf, and one careless intern:

Suppose the files are labelled A and B. As demonstrated in figure 2.1, there are only two possible arrangements for this two-file pile:

Arrangement 1: file B on top of file A

Arrangement 2: file A on top of file B

For this particular case of only two files, both these arrangements are essentially ordered. In other words, if your careless intern accidentally rearranges your files, the worst he can do is just to reverse the order. If you find the files are not in the order you wanted, just turn the pile upside down.

Three files, one shelf, and one careless intern:

Now let's imagine that your job is slightly busier, and that you now have three files in the pile, rather than two. Let's suppose these files are labelled A, B and C. Here comes the bad news: even though we have only increased the number of files by 50 percent, there are now *three times as many* possible arrangements. These are shown in figure 2.1, and are listed below:

Arrangement 1: file C on top of file B, on top of file A

Arrangement 2: file B on top of file C, on top of file A

Arrangement 3: file A on top of file C, on top of file B

Arrangement 4: file C on top of file A, on top of file B
Arrangement 5: file B on top of file A, on top of file C
Arrangement 6: file A on top of file B, on top of file C

So going from just two files to three files means that we have gone from two possible arrangements to six. So now your careless intern can accidentally rearrange the pile in *six* ways.

How many arrangements would there be for four, five, or more files? Obviously more, but how many more? It turns out that there is a simple way of working this number out. Suppose we are building a pile out of three files. There is a choice of three possible files to place at the bottom (A, B or C). Pick one – for example, A. This leaves two possible files in the middle. Again pick one – for example, C. Then there is only one file left to go on top – file B. In other words, we have three possibilities at the bottom, multiplied by two in the middle, and one on top. Writing this in terms of math, this gives a total number of possibilities as:

(3 at the bottom) × (2 in the middle) × (1 on top) = $3 \times 2 \times 1 = 6$, as shown in figure 2.1.

More than three files, one shelf, and one careless intern:

For four files – labelled A, B, C and D – we therefore have $4 \times 3 \times 2 \times 1 = 24$ possible arrangements. For five files we will have $5 \times 4 \times 3 \times 2 \times 1 = 120$ possible arrangements. But all this sounds like a pretty unrealistic office. After all, having ten files in a pile isn't uncommon. So let's have a look at what would happen for ten files. Following the same idea as above, you can see that the number of possible arrangements will be $10 \times 9 \times 8 \times 7 \times 6 \times 5 \times 4 \times 3 \times 2 \times 1$ which turns out to be more than three-and-a-half million. And that really is bad news since it means that your careless intern can accidentally rearrange the pile in more than three-and-a-half million ways!

This gives us our first take-home message about collections of objects. The number of things that can happen to a collection of objects – and in particular the number of arrangements of these objects – quickly becomes very large as you increase the number of objects.

Now, imagine that it is the day before you are going on vacation, and your boss hands you a pile of ten files. She tells you that she has spent the whole day organizing the pile into a special order, according to her own priorities. She also tells you that you should start work on them, beginning at the top, as soon as you get back from vacation. So you place them on your desk, and give everyone strict instructions to leave them alone. Off you go on vacation, and duly empty your mind, aided by the occasional cocktail. Once back in the office, you find a note that your intern has left you – "Very sorry, I accidentally knocked over your files while you were away. But I put them back in a neat pile for you, so no harm done!" No harm done? There are more than three-and-a-half-million possible arrangements, and you have forgotten the particular arrangement which your boss carefully prepared before you left on vacation. Now imagine that you try to randomly rearrange them, hoping that the special arrangement will magically appear before your eyes. Here comes the really bad news. Suppose you spend ten seconds on every arrangement – that means you could search six arrangements per minute, and therefore 360 arrangements per hour. But since the exact number of arrangements is actually 3,628,800, it will take you approximately 10,000 hours to search all of them – and 10,000 hours means that it will take you more than one year even if you don't stop to sleep, eat or go to the bathroom. So unless you have a very patient boss, you will likely be out of a job before you find the correct arrangement.

2.2 If things can get worse, they probably will

In the above story, we imagined that your intern accidentally pushed the whole pile over and hence instantly took the pile from maximum order to maximum disorder. In many other situations, things will move between order and disorder in a more gentle way. Imagine that instead of the whole pile being knocked over all in one go, your intern just randomly changed the position of one file each day. Going back to our picture with only three files, you can see that even after only a couple of days the new pile is likely to be quite different from the original one. Admittedly, the more files there are in the pile, the longer this process of order-to-disorder will take – but in the end, disorder rules.

So, our office filing story tells us that it is quite easy to disorder something which is ordered, while it requires a long time and a *lot* of care to reorder something that is disordered. Exactly how long either of these takes will depend on what is doing the disordering or reordering. But one thing is for sure: *there is a natural tendency for something that is ordered to become disordered as time goes by. In contrast, something that is disordered is highly unlikely to order itself without any additional help*. And herein lies our interest in order and disorder. We have already established that a Complex System such as the traffic or a financial market, can spontaneously move from order to disorder and back again. At the same time, we know that a Complex System contains a collection of objects in a similar way to a pile of files. So how is it that a Complex System can move from order to disorder and back again, all by itself, while something simple like a pile of files cannot?

2.3 We need feedback

There must be some magic ingredient that a Complex System has – but a pile of files or a bag of socks does not have – and which therefore enables the Complex System in question to create order out of thin air all by itself. To help us understand the nature of this magic ingredient, we need to do an experiment:

Grab a ruler from your desk. Now try to balance it upright on your desk. Impossible. Now try to balance it on your outstretched open hand. Again impossible . . . *unless you move your hand continually to counteract the wobbling ruler.*

It turns out that this ruler problem is very similar to the file problem that we discussed earlier. The files can easily go from an ordered to a disordered state, just like a momentarily upright ruler on a desk can easily fall over. But to reorder the files, or to help the ruler stay upright, would require a helping hand. In the case of the files, it requires a very kind boss who is willing to come in and redo all the hard work of prioritization. In the case of the ruler, it requires the actions of a skillful handler to help the ruler stay upright.

The examples of the files and the ruler give us the key to understanding why a Complex System is much more complicated than a collection of objects such as files or socks. In other words, these examples give us the clue to help explain exactly what makes a Complex System so complex, as opposed to merely being a bit complicated. So let's think more deeply about these examples: the only reason that you can balance the ruler on your hand as opposed to the desk, is because your eye notices the movements of the ruler, and then feeds this information to your brain which then *feeds back* the information to your hand in the form of a movement. The same holds for the reordering of the pile of files: it would need your boss to *feed back* the original information into the pile that she injected about the various files' priorities. And that is the answer – feedback – which is a term which appeared on our list of key ingredients of a Complex System in chapter 1.

As we will see, feedback can arise in a given system in a variety of different ways. It can be built into the objects themselves – for example, humans have a memory of the past which can affect their decisions in the present. Or it can be information or influence fed into the system from the outside, as in the case of the balancing stick, or the announcement of news in a market. In the case of traffic, it can be the information that a driver obtains from looking at the cars around him, or listening to traffic reports on the radio.

It doesn't matter where it comes from, it is still feedback – and it is feedback which enables order to "kick in" in different ways and at different times. Feedback can create order in a disordered pile of files, and can order a tumbling ruler into its upright position. However it is typically very hard to see such feedback operate at the level of the individual objects in a particular Complex System – hence it can appear to an outside observer that the order appears out of thin air. This is particularly true if the feedback takes the form of information, since information is not a tangible object. Since drivers in traffic and traders in markets continually input and output information about their own and others' actions, we can begin to see why traffic jams and market crashes could possibly appear out of thin air without any apparent cause. The magic ingredient is the feedback of information.

2.4 Life is just a pocket of order

In terms of our order/disorder story, the upright ruler is an ordered state while the ruler tumbling to the ground is a disordered one. As long as you can keep the ruler upright, you have managed to maintain it in an ordered state. However, you cannot keep this up forever. It takes concentration, and that makes you hungry – and you eventually need to eat. In other words, the origin of the order for the upright ruler is feedback, and this feedback requires the input of energy.

Now, let's take this a step further. The energy which we use to create the feedback for the ruler, comes from the food we eat. And all the food we eat can be traced back to plants. This is even true for meat and dairy products – they come from animals who themselves ate plants. So it all comes down to plants – and plants get their energy from that great energy source in the sky: the Sun. In other words: the Sun represents the root cause of the pockets of order that we observe around us. This is actually quite a deep statement, since it means that the Sun is what helps us buck the general trend from order to disorder. This is even true when we build buildings or create other ordered structures using machines and materials such as concrete. Machines are made of metal and run on gasoline: and gasoline, metal and concrete all originate from the natural resources found on Earth – and these in turn owe their existence to the Solar System and hence the Sun.

So the Sun is the root cause of the pockets of order that we see all around us. And these pockets of order are not restricted to inanimate objects such as files or upright rulers. The town where you live is an example of a pocket of order. It contains many people, organized into houses and streets – and all within the single pocket defined by the town's boundaries. Going further, each of us humans is individually a pocket of molecules which all happen to be piled up into a particular region of space, i.e. within the confines of our own particular body.

2.5 Our Universe's bleak future

We have seen that collections of objects such as a pile of files will, *in the absence of any feedback*, tend to become increasingly disordered. Unfortunately, it turns out that the same is true of the Universe as a whole, and everything in it – including us.

Let me explain the background to this horrifying news. All the evidence gathered so far by scientists suggests that the Universe is isolated. It doesn't touch anything and nothing touches it. Most importantly, there is no feedback of any kind from other Universes – hence there is no "invisible hand" to help keep it ordered. In technical jargon, the Universe is a closed system – and unfortunately there is a fundamental law of physics which states that: *The amount of disorder in a closed system increases as time goes by.* So can we use this law as an excuse for an untidy office? Yes and no. Based on what we saw with the files, we can certainly see why it would make sense in an office setting. However this law is only strictly true for closed systems – and truly closed systems are very rare. In fact, the Universe is the only truly closed system that we know of.

This law tells us that no matter how hard we might try to stop it from happening, the Universe as a whole is heading toward total disorder. In other words, all the objects in the Universe – which ultimately are just collections of molecules – are heading toward total disorder. In short, the future is just one big messy soup of molecules. Now, I am sure that someone is thinking "I am made of molecules. So does it also include me and my molecules as well?" Unfortunately, yes it does. Indeed the phrase "ashes to ashes and dust to dust" captures the whole degenerative process very well. We will all eventually die and our bodies will then gradually decay, decomposing into various pieces and ultimately constituent molecules. Our molecules will then eventually find themselves spread out over the entire surface of the Earth, and then ultimately throughout the entire Universe as the Earth itself disintegrates and the Universe continues its unstoppable march toward increased disorder.

Wait a minute though – we can order files, and we can order the state of a ruler. So doesn't the fact that we can create such pockets of order mean that order has actually increased, and hence that this fundamental law of physics is wrong, and hence we are saved? Unfortunately not. We are certainly able to create temporary pockets of order in certain places and at certain times, if we feed in the right amounts of energy and effort from the outside. However it turns out that this local increase in order comes at the expense of a decrease in the amount of order in your body and in your immediate environment. As you reorder the files or make the ruler stand upright, for example, you are using energy – and some of this energy is lost as heat since you are effectively doing some exercise. And adding heat to your environment means that you are increasing the disorder in the air molecules around your body. In fact it is even worse than this – the disorder which you create as a by-product of your reordering of files or balancing of rulers will always be greater than the amount of order which you manage to create. In other words, the law is correct in that the overall disorder in the Universe increases. So although we humans can invent stories, build buildings, and can even create new lives by giving birth, each of these acts will actually destroy more order in the rest of the Universe than it can possibly create in the resulting book, building or baby.

Depressing? Actually it was a physicist called Ludwig Boltzmann who came up with the pioneering insights into this effect of increasing disorder – and he ended up committing suicide in 1906 by hanging himself while on vacation.

2.6 Air, air everywhere – we hope

But we shouldn't be too glum. It turns out that some disorder is good for us. In fact it is doing us all heaps of good right at this very moment, at various levels of human biology. In particular, it is helping us all breathe easier.

Imagine yourself back in your office, happily getting your breath back after the filing fiasco. You are breathing in air molecules, without a second thought that there may be a sudden shortage of them in the vicinity of your nose. But should we really be taking our next breath of air for granted? It turns out that disorder is our

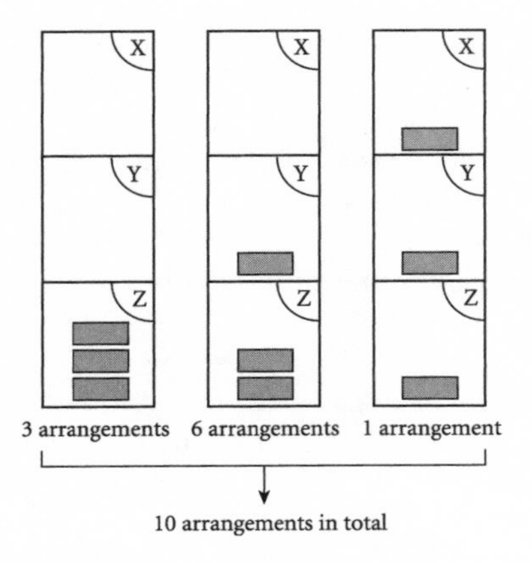

Figure 2.2 Number of possible arrangements of three identical files in a filing cabinet containing three shelves X, Y and Z

savior – and we can understand this in terms of our analogy with piles of files, where the files now represent air molecules. In our filing scenario, we had originally arranged a set of files A, B, C, etc. in a pile on a single shelf. Now let's generalize things slightly, and imagine instead that there are three possible shelves where the files could be placed, with each shelf being the in-tray for a particular employee. Our three employees are Ms. X, Mr. Y and Mrs. Z. If there was only one file, say file A, there would be three possible shelves upon which it could be placed. In other words, there would be three possible arrangements of one file among the three shelves. Specifically, file A could be on shelf X, on shelf Y, or on shelf Z. But imagine that there are three files. Since air molecules are all the same, and it is air that we are trying to understand, we will consider the case where the three files are all the same. So let's think about putting them somewhere on the three shelves.

Figure 2.2 shows that there are ten ways of arranging the three identical files among three shelves. Looking at the left-hand diagram in figure 2.2 where all three files are on the same shelf, there are three possible places for this pile of three files: all three files on shelf X, all three on shelf Y, or all three on shelf Z. Hence there are three arrangements. Things are a little bit more tricky in the

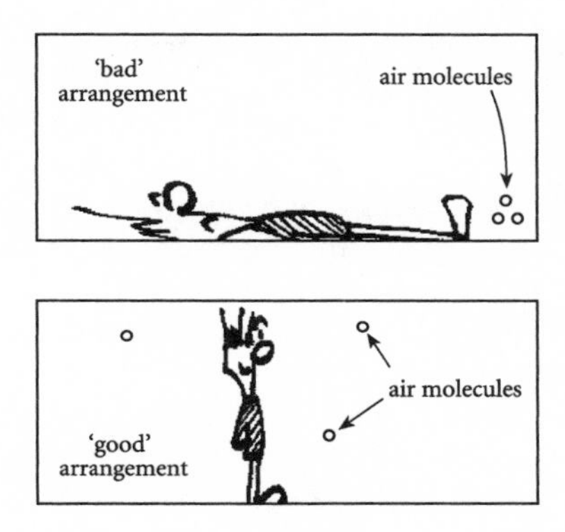

Figure 2.3 Air, air everywhere? Top shows an arrangement where the air molecules are all piled in one region of the room, like having a set of files being stuck on one particular shelf. This "bad" arrangement is very rare. Bottom shows an arrangement of air molecules where they are spread throughout the room, like having a set of files being spread out over many different shelves. This "good" arrangement is very common.

middle diagram, since the pile of two files could go in any of three shelves while the remaining file will have the possibility of either of the other two shelves. Hence there are $3 \times 2 = 6$ possibilities. The right-hand diagram is easier: there is only one way of arranging the three files so that each shelf has just one file.

Suppose that you are lucky enough to be off on vacation again, and that your boss has purposely distributed the three identical files A, B and C, in a particular arrangement among the various employee shelves X, Y and Z. Each day, your careless summer intern comes into your office, takes out a file, and then replaces it at random among the three shelves. Before long, your boss's original arrangement is lost and disorder once again rules. Let's focus on the left-hand diagram of figure 2.2. This tells us that there are certain arrangements which correspond to all the files being in the same place at the same time. If we imagine that the files are air molecules, and that the filing cabinet represents the office itself, this particular arrangement shown in the left-hand side of figure 2.2 would correspond to all the molecules being piled up in one region of the office. So if you happened to be in another part of the office, as in the top of figure 2.3, there would be no molecules within sniffing distance of your nose – and that of course would spell trouble for air-breathers such as ourselves.

So, given that such bad (i.e. unhealthy) arrangements can arise, why aren't we all dropping like flies? The clue to the answer lies within figure 2.2, which shows that there are relatively few arrangements where the objects will all be piled up in one region. Hence the chances of an arrangement arising in which all the air molecules in a room are piled up far away from you is remote. It could happen, but it certainly won't happen very often given the fact that there will always be many billions of air molecules in a room. In fact, I have never heard of anybody experiencing such an effect. In short, disorder saves the day – it guarantees our next breath regardless of where we happen to stand.

But what would happen if an evil scientist did manage to momentarily arrange all the air molecules into one part of the room – would we then die? No, we would hardly even notice it. Air molecules have energy and hence move around and bounce off each other. Therefore it wouldn't be long before all the possible arrangements had been explored – just as with the careless intern and the files. In physics-jargon, we say that the system explores its **state-space** very quickly. So thanks to the very rapid way in which air molecules disorder themselves and the fact that molecules have no intrinsic feedback effect and therefore cannot easily reorder themselves, we never have to worry where our next breath of air is coming from.

I do however feel duty-bound to add a slightly scary caveat. If, for some bizarre reason, air molecules were ever to gain the ability to process information like drivers and traders, and hence introduce feedback into the system, we might see spontaneous order appear in a roomful of air just as it does for traffic jams and market crashes. And then we really would have to worry.

2.7 Our biased world

Complex Systems tend to be "open". In other words, the system interacts with what is around it. But it turns out that the *way* in which it interacts with its surroundings can actually bias the frequency with which particular arrangements of its constituent objects are observed. We humans again provide a good example of this. Although the molecules in our body could in principle start flying around all over the place, they stay within the confines of our body as a result of the interactions that they experience with each other and with the outside world. By eating, drinking and sleeping, we each maintain our body in an alive state – hence we keep our bodies from decaying and thereby keep our body's molecules on the "order" side of the order/disorder divide.

This tells us that the external conditions that a Complex System such as our body experiences can play a major role in biasing the arrangements of objects that are then seen. In the case of our own bodies, this means that the only arrangements of molecules that we observe are the ones in which our own molecules lie within the confines of our own body. A similar situation could arise with our filing problem from figure 2.2: in other words, we could in principle bias the arrangements such that the three files always remain in the same place if we were to put in enough effort and energy.

The fact that biases in the arrangements of objects can arise as a result of external conditions is very important for our understanding of which emergent phenomena are likely to arise in a given Complex System. This is because such biases directly affect which arrangements arise more frequently, and hence are more likely to be observed. Likewise, such biases can also prevent some arrangements from ever occurring. By developing an understanding of the biases introduced by external conditions, we should therefore be able to improve our chances of accurately predicting the system's future behavior. For example, the decision to close off a particular road can dramatically change the frequencies and locations at which traffic jams appear in a particular road network. So given its potential importance for understanding the future evolution of a Complex System, let us explore this biasing effect further using our filing problem:

Three files, three shelves, and one careless intern:

When we discussed figure 2.2, we assumed that employees X, Y and Z were the same in that they were all equally likely to receive a given file. This would be the case, for example, if they all had the same job and worked for the same number of hours. But now suppose that their work contracts are such that X works less hours than Y, who in turn works less hours than Z. Hence X will get less work and hence less files than Y, who will in turn get less files than Z. As a result, certain arrangements of files should arise more often than others. Going back to figure 2.2, we can see that the particular arrangement shown in the middle is probably the most likely. By contrast, if we had said that X and Y were each employed for only three hours per week, but Z was full-time, then the particular arrangement shown on the left would be more likely.

In the above examples, the external condition imposed by the three employees' contracts drives the system toward certain arrangements and away from others. In other words, we have shown how biases in arrangements of objects can arise as a result of external conditions. There are of course many other ways in which the external conditions can bias the likely arrangements. Imagine that there is a government or trade union rule which says that no employee can handle more than one file per week. The only arrangement that should then be seen is the one on the right-hand side in figure 2.2. Another way in which arrangements can be biased arises when the external constraint corresponds to a rule which restricts how some particular subset of the objects can be arranged. In physics-jargon, this effect is called frustration. Now if you are one of the unlucky ones who has to put up with complicated office dynamics, this effect sounds pretty close to home. But it also turns out to be quite a general emergent phenomena in real-world Complex Systems – in particular, the ones which involve collections of objects competing for some kind of limited resource.

To help understand how frustration can arise in a Complex System, we go back to our files. Suppose that for some reason there is a rule that says that file A must not be placed next to C. This may sound unusual for an office but it often arises in a social setting. For example, teachers know that they cannot sit certain kids together because it will cause trouble – and the same can be true at dinner parties. The arrangement of the shelves themselves then becomes crucial. If they are in a line, the problem can be

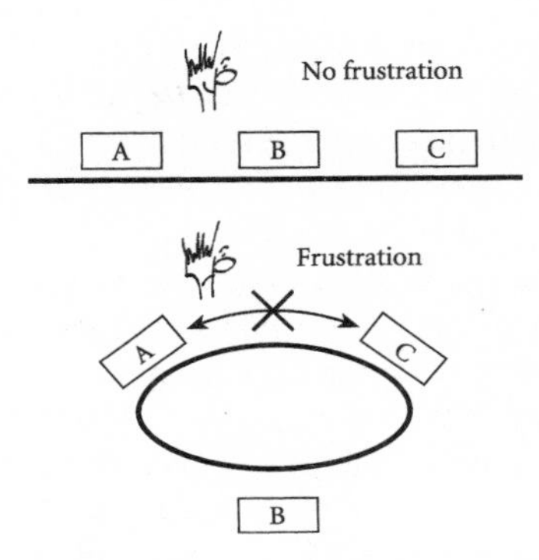

Figure 2.4 The dinner-party dilemma involving three people A, B and C, in which A and C do not want to sit together. Top: a rectangular table allows for a non-frustrated, relatively happy outcome. Bottom: a circular table always leads to frustration.

solved as shown in the top of figure 2.4. But if they are in a circle, then it is impossible. In other words the arrangement is always frustrated as shown in the bottom of figure 2.4.

In addition to unhappy offices and dinner parties, frustration is also quite a common occurrence in the systems that physicists look at since there are certain types of particles – in technical jargon, particles with a certain type of spin – which don't like to be next to each other. Stated more formally, such arrangements may have an unfavourably high energy.

While we are on the topic of physicists, it is worth noting *the* most important example of such biasing which occurs in the domain of Physics – the biasing of arrangements produced by the temperature at which a particular physical system is maintained. In fact the biasing due to temperature is so important to physicists that they have developed a huge amount of mathematical machinery for dealing with it, and have applied this machinery very successfully to many different types of laboratory system. Indeed this approach has been so successful in the Physics domain that many physicists have started trying to apply it to social systems. In particular, there are research papers written by physicists which talk about the "temperature" of a financial market. The problem is that physicists tend to take the specific biasing that temperature causes too seriously. Just because this biasing is correct for physical systems like collections of molecules does not mean that it is also realistic for social ones. Indeed, as we have already discussed, a collection of inanimate objects tends to lack a key ingredient of a social system: feedback. As a result, it is unclear whether the conclusions from physical systems will have much relevance to Complex Systems in a biological or social setting.

But what exactly is the biasing caused by temperature and why is it such a big deal for physicists? In terms of our filing story, the biasing effect of temperature is analogous to that of a tired secretary who is getting ready to leave the office at night and who therefore only has a finite amount of energy available. This secretary will tend to put more files in the lower shelves than the upper ones, in order to save reaching up or having to stand on something. Likewise the amount of energy available to a physical system such as a collection of molecules, is restricted by its temperature. Continuing with the filing analogy, an outside observer who is checking the office's filing arrangement night after night would be left with the impression that the filing system was reasonably well ordered, since the arrangements he observed would tend to be those shown in the middle and left-hand side of figure 2.2. Within the context of physics, the temperature controls the amount of energy available for arranging objects, and this in turn biases the arrangements. As the temperature increases, the amount of available energy increases, and so the biasing becomes less apparent. For the filing analogy, our office observer would then get the impression that the filing system was becoming less ordered since a wider range of arrangements would be observed over time. Eventually, at very high temperatures, the amount of energy available is huge – which is analogous to saying that the secretary has so much energy that she doesn't bias the distribution of files among the shelves in any way. Since the filing process is now unbiased, our outside observer would conclude that the disorder is large since he would observe all possible arrangements of files among shelves with equal frequency.

Hence increasing the temperature in a physical system such as a collection of molecules, generally takes the system from an ordered to a disordered state. The way in which water passes from ice at low temperatures to steam at high temperatures is a great example of this effect. Ice is a solid containing an ordered array of water molecules, while steam is a completely disordered gas. Physicists call the transitions between these different states of water, phase transitions – and the particular mathematical formula that they use to describe how the temperature biases arrangements of molecules is called an exponential or Boltzmann weighting factor. It seems that the popular press on Complexity likes to borrow this type of physics terminology related to phase transitions. However such literal translations of models and ideas from physics should be handled with care, since the biasing in arrangements caused by temperature is only strictly valid for systems such as a collection of molecules which sit in a particular type of laboratory environment. In short, physics has a remarkably large number of answers for certain types of systems – but it is still a long way from having all the answers for general Complex Systems.

Chapter 3

Chaos and all that jazz

3.1 Dealing with office dynamics

In the popular science literature, one often sees the term Complexity bundled together with another "C" word, Chaos. This might suggest that Complexity and Chaos are essentially the same thing. But they are not.

A Complex System tends to move between different types of arrangements in such a way that pockets of order are created – for example, the appearance and subsequent disappearance of a market crash. But we haven't yet said anything about *when* such transitions might occur. In short, we are missing a discussion about time, or what is technically called the dynamics of the system. Given that a Complex System comprises a collection of interacting objects (for example, traders in a financial market), it is likely to exhibit quite complicated dynamics. In other words, the output of the Complex System as seen from the outside by any one of us will appear quite complicated. This word "output" just means any kind of observable number that is produced by the collection of objects. For example, the output of a financial market at any given moment is the price, e.g. $2.50 for a given stock. The fact that the output of a financial market (i.e. price) changes in time in such a complicated way is the reason that we, as outside observers, always see such complicated price-charts appearing in the news for stocks and currency exchange rates.

The way in which the output of a Complex System changes over time falls under the general heading of non-linear dynamics. And Chaos is just *one* particular example of such non-linear dynamics. In fact, the word "Chaos" is used when the system's output varies so erratically that it seems random. The upshot of this statement is that those erratic-looking financial market price charts that we see in the news *could* show Chaos – but they don't *have* to.

Let's try to get to the bottom of all this talk of dynamics, Chaos and randomness, by going back to the office. As some of us know only too well, the way in which people in an office interact can dramatically affect the whole dynamics of the place, and hence can determine what happens to the office itself over time. The same is true for any Complex System. The way in which the constituent objects interact will affect the arrangements that they exhibit, how long they exhibit them for, and the transitions between these arrangements – and this will in turn affect the output of the system such as the price of a given stock. As we will see in chapters 4 and 6, the word "arrangement" in the context of a human system such as a financial market has to do with how the traders arrange themselves with respect to their possible trading strategies. This then determines whether they choose to buy or sell at a given moment – and this in turn gives rise to the price, or output, of the system. But it doesn't matter whether it is files on shelves, or traders with trading strategies – it all still comes back down to a discussion of arrangements of objects.

If there are many possible arrangements of the system's constituent objects, and the system moves in a complicated way between these arrangements, then the resulting output of the system can look random and unpredictable. It is under these conditions that the system might actually exhibit Chaos. If, instead, there is an obvious method to the madness, then the system can look ordered and predictable. The system will then not exhibit Chaos. As our filing story in chapter 2 suggests, the presence of some kind of consistency or memory in the system can be crucial in determining whether the resulting evolution looks unpredictable or predictable, and hence whether it is likely to be Chaotic or not. So, given our interest in predictability, let's now focus in detail on the effect of the intern in the office filing analogy. In particular we wish to compare the dynamics generated by a systematic intern to the dynamics generated by a careless intern, in order to understand how the system's output is affected by the way in which the underlying arrangements change. This will then help us understand the conditions under which a Complex System would be likely to exhibit Chaos. It will also help us understand the conditions under which a Complex System might be predictable.

3.2 Systematic interns, and careless ones

Let's start by considering the following setup.

One file, two shelves, and a systematic intern:

We will label the shelves as 0 and 1, as shown in figure 3.1. In the event that the file is on shelf 0, we will call this arrangement '0'. In the event that the file is on shelf 1, we will call this arrangement '1'.

Suppose that the file starts off on shelf 0, and imagine that our systematic intern has decided that he will, every day, enter the office and change the shelf on which the file sits. Hence the file changes from shelf 0 to 1 to 0 to 1, etc., which we can write as the sequence 0 1 0 1, etc. In other words, anyone noting down the file's location at the end of each day would end up with the following sequence over a period of ten days:

0 1 0 1 0 1 0 1 0 1 . . .

This sequence of observations is referred to as the time-series of the system's output, or simply the *output time-series*. The price-chart in

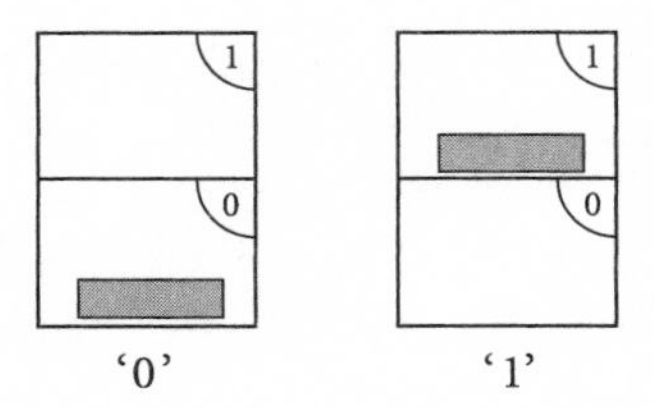

Figure 3.1 The two possible arrangements for one file and two shelves

a financial market is another such time-series. For example, such a price-chart could conceivably look something like this:

> $2.37 on day 0, $2.34 on day 1, $2.65 on day 2, $2.44 on day 3, $2.48 on day 4, $2.34 on day 5, $2.43 on day 6, $2.32 on day 7, $2.48 on day 8, $2.35 on day 9, $2.46 on day 10

which can be written more simply as

> $2.37 $2.34 $2.65 $2.44 $2.48 $2.34 $2.43 $2.32 $2.48 $2.35 $2.46

This certainly looks like a lot of information. In fact, the detailed time-series that emerge from Complex Systems like financial markets can literally represent too much information for us humans to process. Many of us are really only able to take in whether the output (price) has gone up or down. In other words, instead of a long list of numbers we would instead think of a long list of ups and downs. If we write an "up" as 1 and a "down" as 0, then this would give a list of 1's and 0's. In our example above, the price goes *down* from $2.37 on day 0 to $2.34 on day 1 – hence this is a 0. By contrast, the price goes *up* from $2.34 on day 1 to $2.65 on day 2, which is a 1. Working out the changes in price from day to day in our example gives the following simplified version of the financial market price-series:

> 0 1 0 1 0 1 0 1 0 1 . . .

which happens to be the same time-series as the filing example above.

Now let's imagine that we, as outside observers, turn up at the office or market and are faced with such a time-series of 0's and 1's. This time-series tells us what has happened in the recent past. Our job as office observers, or as market speculators, is to work out what will happen next. Just to remind ourselves, the time-series looks like

> 0 1 0 1 0 1 0 1 0 1 . . .

and hence looks very ordered in time. Even if we don't know the systematic rule which the intern is using to move the file in the office scenario, the fact that this time-series is so ordered means that we could probably guess the rule ourselves and even make a reasonably accurate prediction about the next outcome – which we would probably all guess would be a '0' in this example. The same idea holds for the scenario of a financial market – in other words, whether the price moves up or down looks to be predictable in our example, even though the actual value of the price might not. You might think we would be pretty lucky to find such an ordered pattern in a financial market – and you would be right, partially. For example it turns out that the dollar-yen exchange rate did, for several years during the 1990s, indeed have such an underlying pattern.

So, for an office with two shelves, it might be reasonably straightforward to make a prediction as long as the intern follows a systematic rule for changing the file's position. Likewise for a financial market with an up-down price-series, it might be possible to make a prediction of future up-down movements as long as the collection of traders is operating as a crowd, and can hence generate an ordered series of price movements. We return to this point in chapter 6.

Now imagine replacing the systematic intern by a careless one, who moves the file around randomly. In other words,

One file, two shelves, and a careless intern:

This is equivalent to saying that he just flips a coin each day in order to determine the file's new position, with heads representing '1' and tails representing '0'. The time-series will now become completely random, with typical sequences such as

0 1 0 0 0 1 1 0 1 0 . . .

which contain no patterns whatsoever. In other words, the time-series looks completely *disordered in time*. If we were outside observers we would now say that the series looks unpredictable. The same holds for a financial market: if the traders are not acting as a crowd, the price-series is much more likely to resemble this random one shown above, as opposed to the ordered one shown earlier.

3.3 Don't worry, it's just chaos

In our two-shelf example, the systematic intern gave rise to a series of outcomes which are highly ordered in time, and hence a predictable time-series, while the careless one gave rise to a series of outcomes which are highly disordered in time, and hence an unpredictable time-series. This could be turned into a nice neat take-home message about predictability were it not for the phenomenon of Chaos.

It turns out that even a systematic intern can, if he uses a sufficiently complicated rule and if the number of possible arrangements is sufficiently large, produce a time-series which looks highly disordered in time, and hence unpredictable. Strictly speaking, chaotic time-series do have a predictable pattern in them. However, it is so hard to find it that it might as well not be there – and therein lies the problem with trying to predict the dynamics. So let's see how this works by considering a setup in which the file-moving rule is complicated and where there are many possible arrangements.

One file, many shelves, and a systematic intern:

We consider the case of many shelves, and hence many possible arrangements of the one file. The intern uses the following systematic rule for determining the file's next position. Since this rule is quite complicated to explain, we will write it as a list of instructions:

Step 1. Calculate a number S which is given by the shelf number on which the file is located divided by the total number of shelves. In other words, S is a number between 0 and 1 which expresses how far up the filing cabinet the file is. So $S = 1$ means the file is on the top shelf and $S = 0$ means the file is on the bottom shelf. $S = 0.5$ means the file is half-way up, and would correspond to the file sitting on shelf 50 in a filing cabinet containing 100 shelves for example. $S = 0.25$ means the file is a quarter of the way up, and would correspond to the file sitting on shelf 25 in a filing cabinet containing 100 shelves.

Step 2. Suppose the file starts off sitting on a given shelf, and that this corresponds to a particular value of S which we will refer to as S_1. In order to work out which shelf to move the file to, which we call S_2, the intern takes S_1 and multiplies it first by $(1 - S_1)$ and then by a number r. Let's choose $r = 4$, and $S_1 = 0.4$. This means that $(1 - S_1) = 1 - 0.4 = 0.6$ and hence the new shelf location S_2 is given by

$$S_2 = 4 \times 0.4 \times 0.6 = 0.96$$

In mathematical terms, the formula which the intern has used – and which by the way is the only formula in this book – is given by

$$S_2 = r \times S_1 \times (1 - S_1)$$

Step 3. The intern now repeats Step 2, but with S_1 replaced by S_2 and S_2 replaced by S_3. In other words, he uses the formula

$$S_3 = r \times S_2 \times (1 - S_2)$$

and hence obtains $S_3 = 4 \times 0.96 \times (1 - 0.96)$ which means that $S_3 = 0.15$.

Step 4. The intern repeats this process over and over again in order to obtain all subsequent shelf locations. In other words, he obtains S_4 from S_3, then S_5 from S_4 and so on.

If you were to follow this set of instructions expecting the number S to eventually settle down to some particular value, you would be in for a surprise – it never does. Not only that, but there is no discernible pattern at all. This is because you have created a time-series which is chaotic. In other words, you have uncovered Chaos. Now, maybe you wouldn't have expected S to ever settle down and therefore you feel like I have wasted your time. If this is the case, let me quickly re-surprise you. Go back and repeat the whole process, but instead of using $r = 4$ you now use any value of r between 0 and 1. For example, let's choose $r = 0.1$, and let's still use $S_1 = 0.4$ as above. The new shelf location S_2 is given by

$$S_2 = 0.1 \times 0.4 \times 0.6 = 0.024$$

and hence the next shelf location S_3 is given by

$$S_3 = 0.1 \times 0.024 \times 0.976 = 0.0023$$

Keep going with this and you will find shelf locations S that get closer and closer to zero, i.e. the file moves quickly to the bottom of the filing cabinet, which corresponds to $S = 0$, and stays there. It is as though the file has been attracted to a particular point in the filing cabinet where it then becomes fixed for all time. We have just uncovered a so-called fixed-point attractor of the system's dynamics. By contrast, the earlier example with $r = 4$ was very strange in that the file didn't seem to be attracted to any particular shelf – in fact, the value of S never repeats itself. The technical term for this type of behavior is, somewhat unsurprisingly, a strange attractor.

Let's just take a moment to catch our breath and think through the implications. A systematic intern applying the complicated rule that we wrote down with $r = 4$, and with a filing cabinet with many shelves, will produce Chaos. Although the successive locations, i.e. successive S values, looked like they occurred randomly, this is only because the rule was so complicated that it produced a very complicated output time-series. There was still method in the madness, in that the systematic intern knew exactly what he was doing. And unlike a careless intern flipping a coin, he would get the same result on a given day no matter how many times he repeated the calculation. So the file would always end up on the same shelf on a given day. In terms of us as outside observers just looking at this output time-series, we could *if we were really clever* deduce the rule that the systematic intern used – and hence make an accurate prediction about the file's next location – just by observing the shelf positions over many, many days. In other words, a rule exists and it would be up to us to find it. Actually, I am not sure that I would be able to find it in practice, but at least it is possible in principle – and that is good to know. Indeed, this is very much like when you take over someone's job in an office if they are suddenly away on sick leave – you know there must be some kind of logic to their filing system, but it can really take a lot of effort to finally work it out.

This filing example has also shown us something else – something rather curious, and ultimately very worrying. The same systematic intern, using the same rule but just changing very slightly the value of the number r from $r = 4$ to a number between 0 and 1 ($r = 0.1$ in our case), managed to completely change the time-series of file locations that was produced. Instead of bouncing around all over the place, the file moved quickly to the bottom of the filing cabinet ($S = 0$) and then just stayed there. And this type of behavior is completely the opposite of Chaos. So we have uncovered an important take-home message for understanding the types of behaviors that a Complex System can show. Even if a system has the same setup – in our case, the same systematic intern, the same rule for changing shelf, the same number of files and the same number of shelves – there can be a wide range of outputs, or in other words a wide range of dynamical behaviors. One such example is Chaos, but there are others.

Maybe you are thinking that the consequences for understanding a given Complex System aren't actually that bad. Maybe things are either chaotic, and hence the time-series of outputs appears disordered in time and therefore essentially random (e.g. $r = 4$), or they are completely ordered in time (e.g. $r = 0.1$). Not quite, unfortunately. It turns out that the road between these two is also complicated – in other words the route to chaos is quite a rocky road. And in our case, we can take ourselves along this route simply by changing the value of the number r. Specifically, we can take the system from the regime where the output time-series is ordered in time, with r between 0 and 1, to the regime where the output time-series appears to be disordered in time, with $r = 4$, just by changing the value of the number r. As we will see, the variety of possible behaviors that we uncover is enormous. And given that our intern-filing scenario is an extremely simple example of what a Complex System can actually do, it follows that any given real-world Complex System might show an equally wide range of such behaviors – and maybe even more. So let's investigate more carefully this panorama.

With r between 0 and 1, the output time-series is extremely ordered in time. No matter where the file starts off, it quickly makes its way down to $S = 0$, and then just stays there. Let's suppose that the systematic intern now chooses r to be bigger than 1, for example $r = 2$. Starting again with 0.4, the sequence of successive shelf locations is:

0.4 0.48 0.5 0.5 0.5 0.5 . . .

So instead of heading toward the bottom of the filing cabinet, the file heads toward the middle and stays there. Now let's suppose that the systematic intern had chosen a value slightly larger than $r = 3$, such as $r = 3.2$. In this case, the resulting time-series of file locations eventually repeats itself – in particular, the following pattern emerges:

. . . 0.80 0.51 0.80 0.51 0.80 . . .

In technical jargon, the time-series has become periodic and hence repeats itself after every two steps. It therefore has a period equal to two. This is very strange since there is nothing in the rule which the systematic intern uses which would suggest that the file should move between two shelves in such an ordered way. And yet the file just bounces back and forth between these two shelves like the tick-tock of a reliable clock. Amazing – but things get even stranger on our route to chaos as r increases toward $r = 4$.

Suppose that the systematic intern chooses a slightly larger value of r, such as $r = 3.5$. The resulting time-series suddenly stops repeating itself after every two steps, and instead repeats itself after every four steps. It has a period of four. So the file will move between four shelves in an ordered way, going back to the same shelf every four steps. Yet all the systematic intern did was to change slightly the number r in the rule that he was using.

Further increasing r toward $r = 3.6$ gives a time-series of period eight, then sixteen, then thirty-two. In fact it keeps doubling in this way until the period is so long that it looks like it never repeats itself. And this is just like the Chaos that we saw earlier for $r = 4$. In fact, Chaos can simply be seen as a periodic pattern whose period is so long, that the pattern never repeats itself. Now that really is remarkable behavior, by anybody's standards.

We can represent all this with a special type of diagram which shows the final shelf locations, i.e. the S values, for different values of r. Since it is easy to work out these final S values as long as you have a calculator lying around, we might as well go ahead and plot the final S values for all r values at the same time. The result is shown schematically in figure 3.2 and what it means is

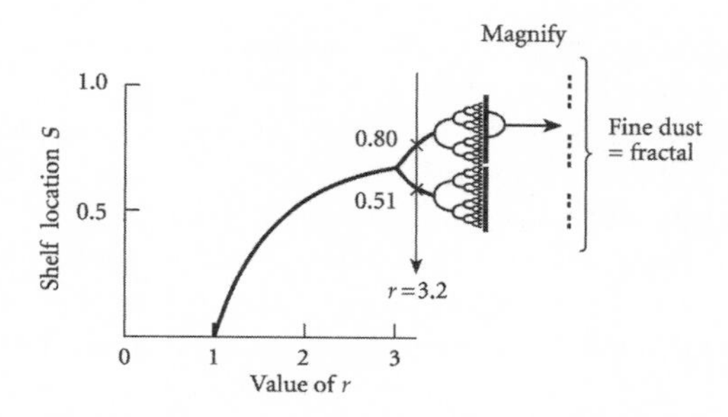

Figure 3.2 A schematic diagram showing the shelf location where the file ends up after many steps, for a range of *r* values. As the *r* value increases above 3, the number of locations that the file ends up moving between doubles rapidly. In the region of $r = 3.6$, which is shown magnified, the number of locations becomes so large that the value of *S* never seems to repeat itself. For this reason, the pattern ends up looking like a solid line. But it isn't solid. Instead it is like an extremely fine dust containing many, many points. It is called a "fractal". For *r* values from 3.6 to 4 (not shown) the dynamics remain chaotic, apart from occasional glimpses of periodic behavior.

this: pick a value for *r*, and find it on the horizontal line. Then look straight up vertically to read off the corresponding values of the black lines. These are the final shelf locations, i.e. the *S* values, between which the file will end up bouncing forever – these shelves, and no others. So, for example, take the case $r = 3.2$ which we considered earlier. Find this value on the horizontal line in the diagram. Then if you look straight up, you will find the two values 0.51 and 0.80 which are the two values which keep repeating themselves in the time-series:

. . . 0.80 0.51 0.80 0.51 0.80 . . .

For values of *r* near 3.6, the file ends up perpetually moving between different locations with the associated value of *S* never repeating itself. As a result, there seem to be so many points on the diagram that it looks like a solid vertical line. But it isn't – instead it is like a very fine dust of points. And here comes something rather peculiar: there are actually an infinite number of points, and an infinite number of gaps. So this apparent line, which is shown magnified on the right-hand side of figure 3.2, is between being an infinite set of points and a solid line. Now, it turns out that scientists refer to a point as being zero-dimensional, a line as being one-dimensional, and a flat sheet such as a television screen as being two-dimensional. This fine dust of points which looks like a solid line but isn't, is effectively between a point and a line – hence it is between a zero-dimensional object and a one-dimensional object. As we all know, a number between zero and one is called a fraction – hence the fine dust of points has a fractional dimension. For this reason, scientists call an object such as this fine dust of points, a fractal.

You can also see a repeated pitchfork shape emerge as the value of r increases toward 3.6. Each line splits into two – and this repeats itself on an ever smaller scale. It is equivalent to saying that the period keeps doubling from 2 to 4 to 8 etc. As we noted above, the final object is a line of points – like a dust of points – which would look to your eye as you move toward the page as "points within points within points". Here we have the same thing: a pitchfork-like pattern within a pitchfork-like pattern within a pitchfork-like pattern. This pattern-within-a-pattern repeated over and over is again referred to as a fractal.

This emergence of fractals is a common occurrence in Complex Systems, both in terms of the output which a Complex System produces in time and the resulting shapes which emerge in space. In other words, fractals are a typical emergent phenomenon of Complex Systems. Just as with Chaos, this does not mean that fractals are always observed in a Complex System – just that they can be. Given this widespread interest in fractals, the two boxes marked "Fractal Fun" show a couple of ways in which you can generate fractals using just a pen and paper. This is not how real-world Complex Systems actually generate their fractals – far from it, since there are many ways of generating the same fractal. But they do help illustrate what a fractal is.

Suppose that our systematic intern is now dictating the price of some commodity in a market, as opposed to dictating the location of a file. The position of the file S becomes the market price.

Fractal Fun I: Dust to dust

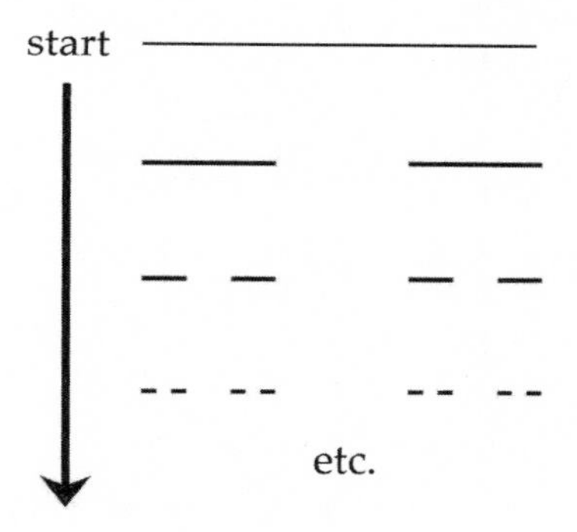

The figure above shows how to produce a dust-like fractal. Start by drawing a horizontal line. Divide it into three, and remove the middle piece. This leaves you with a straight line with a hole in the middle. In fact the easiest way to do this, is to draw the new shape you generate in an empty space below the present shape, as shown. Treating each of the two resulting pieces as a new line, divide each one into three and again remove the middle piece. Keep repeating this process over and over again, for as long as you can. You will end up with a fine dust of points – in other words, a fractal having a fractal dimension between zero and one. This fractal looks very similar to the one that turns up in the regime of Chaos in figure 3.2.

Fractal Fun II: Let it snow, let it snow, let it snow

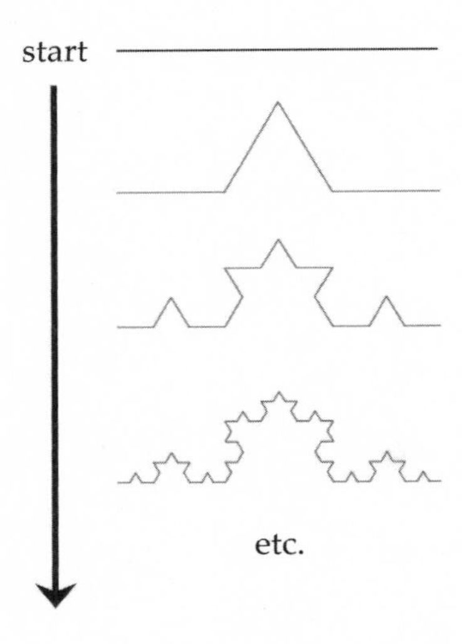

The figure above shows how to produce a snowflake-like fractal. Start by again drawing a horizontal line. Divide it into three, but now replace the middle piece by two pieces which are equal in length to the removed piece. This leaves you with a straight line with a hat shape in the middle, as shown. Treating each of the resulting pieces as a new line, divide each one into three and again replace the middle piece by two pieces of equal length. Keep repeating this over and over again. This time the fractal looks like it contains so may lines that it begins to fill the page. In other words, this snowflake-like structure has a fractal dimension between one and two. Similar shapes can arise in many real-world systems – for example, the border of a cancer tumor.

Instead of a time-series of file locations as we had above, we would have a series of prices – a price time-series. The rich variety of behavior that this price series could then show is indeed consistent with the wide range of behaviors that we see emerging from financial markets – from moments where the price doesn't seem to change (e.g. $r = 0.1$), to moments where it appears to oscillate back and forth in a so-called business cycle (e.g. $r = 3.2$), through to moments where it appears to be random (e.g. $r = 4$). In other words if our intern-turned-price-maker increased the value of r from zero to four, the price would range in behavior from something that never changed over time, to something that repeated itself, to something that looked quite simply chaotic.

Now, if all this can happen with such a simple setup as one systematic intern and a filing cabinet, imagine what could happen in a system containing a collection of such people or objects? The short answer is "all this and more". But don't worry – we haven't been wasting our time. It turns out that the range of behaviors which arise in this simple example tend to be ones that are often observed in real-world Complex Systems. For example, the heart is a Complex System containing a collection of cells which are interacting in a complicated way, with feedback. The result is an output which seems to oscillate, or "beat", in a fairly ordered way – but it can occasionally behave erratically. Meanwhile a market price is typically random looking – but it can occasionally appear more regular with oscillatory-type behavior emerging every so often.

3.4 If I remember correctly, I am living on the edge

We just saw how a systematic intern could generate Chaos, and hence fractals, in the location of a file in a filing cabinet. I also gave you two mathematical rules which had nothing to do with files or interns, but which still managed to generate fractals. Now let's have a look at those fractals again. One could argue that the dust-like fractal shown in figure 3.2 and in "Fractal Fun I", looks a bit like traffic seen from way up in the sky – or even a trail of ants. And the snowflake-like fractal in "Fractal Fun II" could be said to look like the coastline of an island. Its jaggedness might also make you think a bit about mountain ranges, or even those stock price-charts that we mentioned. Because of this apparent similarity, it is precisely at this stage of the discussion of Complexity that many articles and books on the subject tend to stop dead. Here is their reasoning: a systematic rule can produce Chaos; Chaos has fractals in it; fractals look a bit like things we see all around us; so they must all be one and the same thing. Case closed.

But that kind of reasoning does very rough justice to the whole Complexity mystery. Indeed just like an investigation into a crime, where finding one possible motive doesn't mean we have found *the* motive, finding one possible way of generating complex emergent behavior doesn't mean we have found *the* way it is actually generated in real-world Complex Systems. Just take the dust-like fractal, and think of traffic. Fractal patterns are indeed observed occasionally in everyday traffic – however there is no systematic intern organizing the traffic and thereby producing the observed fractal pattern. Nor is the line of drivers on the road magically dividing itself into three, removing the middle section, and hence going through the fractal-generating rule I gave you in the previous section. Even though it is encouraging to have shown that relatively simple rules can give such phenomena, it in no way explains how a Complex System comprising many interacting parts, manages to produce such phenomena. This is precisely the reason why this book won't go the route that so many others have by focusing on different types of mathematical rules for producing non-linear dynamical phenomena such as Chaos and fractals. There is already a branch of science that does that, and people like Steve Strogatz have written excellent articles and books on such machinery and effects.

So, even though the examples in this chapter have helped us understand the range of behaviors that a real-world Complex System can show, they are far from the end of the story. In short, the behavior of real-world Complex Systems is even more complicated than the behavior produced by repeatedly applying a single mathematical rule, for the following reasons:

(1) Real-world Complex Systems contain collections of objects whose complicated overall interactions feature feedback and

memory. By contrast, our intern problem featured a single systematic intern who relentlessly applied the same complicated mathematical rule over and over again.

(2) In order to produce the wide variety of outputs such as Chaos and fractals, the systematic intern had to manually change the value of r. In other words, the intern becomes the "invisible hand" or central controller that is absent from real-world Complex Systems. Instead, real-world Complex Systems can move around between order and disorder of their own accord.

(3) It is inconceivable that any living object – such as a driver or trader, animal or even a living cell – can go through its entire existence taking actions which are dictated by a single, highly-complicated mathematical rule such as the one presented earlier. The complexity observed in real-world systems cannot have been generated simply by applying such a rule over and over again.

Later in the book we will focus on real-world systems in which points (1) and (2) are important. Here we look more closely at point (3). As we all know, real people are not as systematic as our systematic-intern example claimed they could be. In fact they probably have moments when they are systematic and moments when they are not. I know I do – and I know quite a few others that do as well. After all, we are only human. We know from section 3.2 that a completely careless intern will produce a time-series that is random. So what happens if we have a more realistic intern – one who is neither completely systematic nor completely careless? We need to know this, since when we deal with Complex Systems involving a collection of people, there will undoubtedly be such a mix of both things going on. Not only will there be some people behaving more systematically than others at any one moment, but these people's behavior will itself change over time. This is the case that we now consider – the more realistic intern – and this will nicely bridge the gap between completely systematic but complicated behavior such as Chaos, and completely random behavior. In particular, the output time-series will lie between complete order and complete randomness. Not only that, but the type of patterns produced will turn out to be mirrored in a wide range of human activities – from music and art through to human conflict and financial market trading. Moreover these are also the types of patterns that are observed in physical systems as diverse as the sizes and shapes of galaxies through to the coastlines of countries. The upshot is that there is indeed a type of *universal pattern of Life* lying somewhere in the middle-ground between completely ordered patterns and completely disordered patterns. And such patterns are produced by objects that are neither completely systematic nor completely random – in other words, objects like us. So it looks like we really do live in some kind of middle-ground.

So what happens in our filing problem with one file and many shelves, where the intern isn't completely systematic but also isn't completely careless? Since anything not completely systematic will have, by definition, an element of carelessness, our analysis will involve coins being flipped to mimic this carelessness effect. We will keep thinking about one file and many shelves – but instead of letting the intern move the file from any one shelf to any other, we will for simplicity restrict him to moving the file by at most one shelf, either upward or downward. This is perfectly fine for illustrating what we want to illustrate, and could easily be generalized. It also mirrors a lot of real-world systems where the changes from one step to another are reasonably gradual.

Here is our modified story. Each day, our intern will shift the file up or down one shelf – and we will look at what happens as we move from this being done in a completely random way, through to it being done in a systematic way. Let us start off with the completely random way – in other words, we have a careless intern. And the fact that this careless intern is equally likely to move the file up or down one shelf is equivalent to saying that he moves the file according to the toss of a coin. So each day, the intern tosses a coin: if it is heads, he moves the file up one shelf; if it is tails, he moves it down one shelf. Since there are many shelves, and we can assume that the file starts off somewhere near the middle, we won't need to worry about the file reaching the top or bottom. So, if the file starts on shelf 10, for example, the time-series showing its location over consecutive days might look as follows:

10 11 10 9 10 11 12 11 12 13 . . .

In technical jargon, the file is undergoing a *random walk* or a *drunkard's walk* since a drunkard presumably has equal chances of walking forward or backward at each step, like the file. The coin-toss rule that our careless intern applies doesn't have any memory. In other words, the coin is equally likely to show heads or tails on a given day irrespective of the previous outcomes. Coins don't remember anything – and nor, supposedly, do drunkards.

But that may be unfair on drunkards. What happens if the drunkard *does* remember what general direction he is heading in? In other words he has some memory of the past. In the filing context, this corresponds to making the intern more systematic. We can simulate this effect of adding memory, and hence adding feedback to the coin-toss operation, by saying that the likelihood of tossing heads depends on what has happened in the past. Using the technical jargon we have come across earlier in this book, we are gradually adding memory or *feedback* to the dynamics. And what we will find is that this feedback – which we have already stated is a crucial ingredient of Complex Systems – helps create exactly the same kinds of patterns which are observed in the output time-series of many real-world Complex Systems.

The memory, or feedback, that we will add is simple. If the file happens to be moving upward, for example, it will then have an increased chance of continuing to move upward. Likewise if the file happens to be moving downward, it will have an increased chance of continuing to move downward. This is equivalent to saying that the coin is biased in such a way that the random walk now has some memory in it. An example of the resulting time-series showing the location of the file, would be:

10 11 12 11 12 13 14 13 14 15 . . .

Since the earlier time-series without any memory was called a drunkard's walk, we could call this a "slightly sober drunkard's walk" if we wanted. Now if we continue making this slightly sober drunkard's walk increasingly sober – and hence increasingly biased – then eventually the file would move as follows:

10 11 12 13 14 15 16 17 18 19 . . .

and this is exactly the output time-series that a systematic intern would produce if he was moving the file up one shelf every step. It is also the outcome we would get from a completely sober walker walking steadily down the street.

So the random-walk of the file gradually becomes less random, and ends up as a steady walk. But how can we characterize this a bit more scientifically? This opens up one of the hard questions that scientists have – they can often see some kind of order, but need to find a way of characterizing it. Now, each time the intern carries out the experiment with a given bias in the coins, he will get a slightly different answer. Therefore whatever way we find of characterizing the order will need to be a statistical way. In other words, it tells us what is happening on average. More specifically, it is the result of averaging over lots of different experiments with the same setup – so, lots of different offices or lots of different drunkards with the same memory.

It turns out that the way in which scientists typically characterize such walks, relates to the drunkenness of the person doing the walking. Consider the sober walker, who walked as you and I would typically walk down a street. Recall that the position at successive steps is given by:

10 11 12 13 14 15 16 17 18 19 . . .

In other words, after nine steps he has moved a distance equal to nine, i.e. since the final position is 19 and the initial position was 10, he has moved a distance 19–10 = 9 in 9 steps. Equivalently, the file moves 19–10 = 9 shelves in nine steps. This means that the distance moved is nine, in a time interval of nine steps. The distance moved can therefore be written as t^a where $a = 1$ and $t = 9$. (Any number calculated to the power of one is equal to the number itself. Just try it on a calculator.) Note that we could have also referred to the sober walker as creating a perfectly persistent walk, since he always persists in the same direction as he is already going. By contrast, recall the output time-series for the completely drunk walker, or equivalently the file position when being moved by the completely careless intern:

10 11 10 9 10 11 12 11 12 13 . . .

In this case, after nine steps he has moved a distance equal to about three, i.e. since the final position is 13 and the initial position was 10, he has moved a distance 13–10 = 3 in 9 steps. Now, we know that 3 × 3 = 9, or equivalently that three is the square root of nine. In mathematical terms, this can be written as $3 = \sqrt{9}$ or equivalently $3 = 9^{0.5}$. This means that the approximate distance moved can be written as t^a where $a = 0.5$, as opposed to $a = 1$ for the sober walker. In other words, the approximate distance moved can be written as $9^{0.5}$, as opposed to 9^1 for the sober walker.

For intermediate cases where the walker is neither completely sober nor completely drunk – or, equivalently, where the intern is neither completely systematic nor completely careless – the approximate distance moved can be written as t^a where a is larger than 0.5 but smaller than 1. Just look at the output time-series that we had earlier for this case:

10 11 12 11 12 13 14 13 14 15 . . .

The distance moved in $t = 9$ timesteps is 15–10 = 5, which can be written as t^a with a given approximately by 0.74.

So as we span the range from drunk to sober, the corresponding walk goes from covering an approximate distance of t^a with $a = 0.5$, to t^a with $a = 1$. Equivalently, if we span the range from a careless intern to a systematic intern, the change in file position goes from an approximate distance of t^a with $a = 0.5$, to t^a with $a = 1$. In other words, the walk increases in its persistence. By contrast, now imagine we had biased the walk such that there was an anti-persistence in terms of carrying on in one direction. This could be achieved by biasing the coin-tossing so that an outcome of heads would more likely be followed by an outcome of tails. In this case, the approximate distance moved would be given by t^a with a less than 0.5 but greater than $a = 0$. An extreme case would be where the bias was such that every heads is followed by a tails and vice versa – hence every move up by one shelf is followed by a move down by one shelf and vice versa. In this case, the distance moved after nine steps will be just one. Using t^a then gives $a = 0$ since any number calculated to the power of zero is equal to one.

We have therefore found a way of characterizing the order, or conversely the disorder, in an output time-series by calculating the approximate distance moved, and then relating this to the number of timesteps using t^a. This procedure then gives a particular value of a. For technical reasons, some scientists prefer to think about a number D defined by $a = 1/D$. So $a = 0.5$ corresponds to $D = 2$, since $0.5 = 1/2$, and $a = 1$ corresponds to $D = 1$, since $1 = 1/1$. Furthermore, since this is a statistical characterization of how ordered the output time-series is, we could refer to the resulting number D (or equivalently $1/a$) as a statistical dimension. This means that for a value of a between $a = 0.5$ and $a = 1$, the dimension is between $D = 2$ and $D = 1$. And as we saw earlier, since any number between 1 and 2 is a fraction, we could legitimately refer to this as a fractional dimension. In other words, an output time-series with a fractional value of D, is a fractal.

To refer to such a walk as a fractal, makes sense when we think about the shape of the resulting output time-series. Figure 3.3 shows sketches of typical shapes for the cases which we have

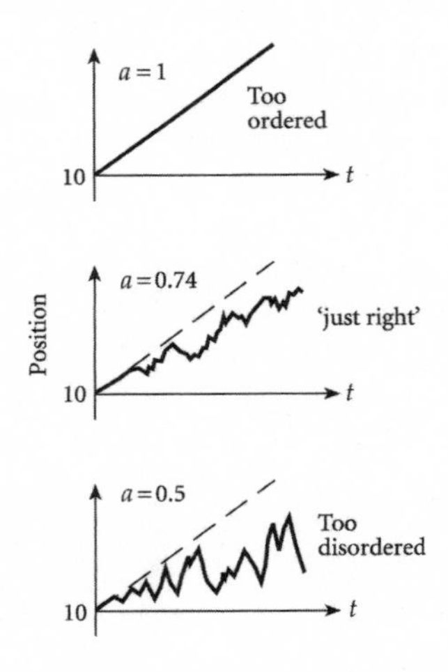

Figure 3.3 Not too ordered, not too disordered, but just right. Most of the shapes which we observe in the real world, are neither too smooth (i.e. too ordered) nor too jagged (i.e. too disordered). In technical jargon, they are referred to as fractals.

discussed: $a = 0.5$, $a = 0.74$ and $a = 1$. Of all the shapes shown, the one that looks closest to that observed, say, in real mountain ranges or coastlines, is the one with $a = 0.74$. In other words it is neither too jagged, nor too smooth. By contrast, the others look too jagged, such as $a = 0.5$, or too smooth, such as $a = 1$. Another way of saying the same thing is that the output time-series for $a = 0.5$ and hence $D = 2$, is too disordered (or too random) for a real mountain range or coastline. On the other hand, the output time-series for $a = 1$ and hence $D = 1$, can be said to look too ordered (or too systematic, or too deterministic).

3.5 There's music, and then there's everything else

Real mountain ranges and coastlines look to be better described by a value of a between 1 and 0.5, and hence a fractional dimension D between 1 and 2. In other words, mountain ranges and coastlines seem to be fractal. And here is the interesting piece of news – so does nearly everything else. More precisely, it does seem to be at least approximately true for a large number of complicated patterns which emerge in our everyday world. In later chapters we will mention specific examples in the economic and sociological domains. But for now, we will just try thinking through the consequences of having such a seemingly "universal" pattern in our lives. The fact that the value of the parameter a lies between 1 and 0.5, and hence the fractional dimension D lies between 1 and 2, means that the output time-series lies somewhere in the murky region between complete order ($a = 1$) and complete disorder ($a = 0.5$). But why should so many other things in our lives also inhabit this murky region?

To answer this, let's think for a moment about music. When many people who are not musicians sit down at a piano, they try to play something simple like *Three Blind Mice*. However, I bet there are not many people who have chosen to download *Three Blind Mice* onto their MP3 player. Why? Because it is a boring tune. In fact if you look at the shape of the music for *Three Blind Mice*, as sketched in figure 3.4, you can see that it is very ordered – too ordered to be interesting, in fact. Now, when we hear a piece of music played on a piano, for example, the shape becomes the

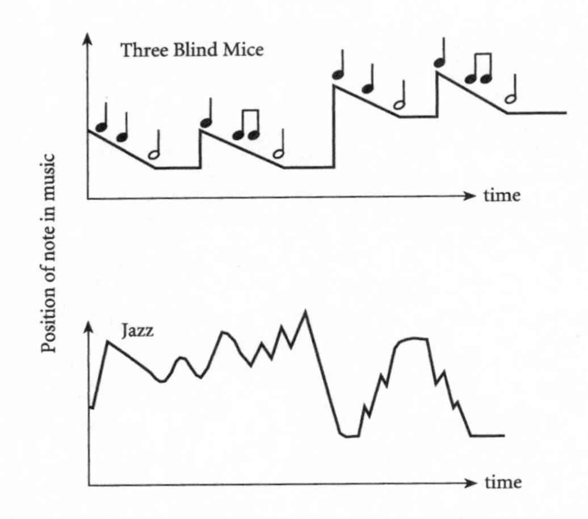

Figure 3.4 The shape of music when written out on music paper, or equivalently when played in time on an instrument. Top diagram shows a very ordered and hence boring piece such as "Three Blind Mice". Bottom shows a less ordered and hence more interesting piece, such as part of a jazz solo in bebop style.

output time-series of the piano – in other words, music can be seen as the output time-series from a system comprising the player and the instrument. Hence an equivalent way of explaining why simple tunes like *Three Blind Mice* seem boring is that the output time-series is too ordered.

By contrast, the music that many people would class as interesting – including classical music by composers such as Bach through to modern jazz – has a far more complicated structure. In other words, it is far less ordered, or equivalently far more "surprising". Indeed such music typically has, when viewed on a page as in figure 3.4, a fractal-like shape containing patterns within patterns. In particular, the bottom diagram in figure 3.4 shows the pattern of one of the saxophone solos of Charlie Parker, who was one of the legends of Bebop and hence modern jazz. It is a much more intricate pattern than *Three Blind Mice*, and contains "patterns within patterns" as we had for the fractal shapes with a between 0.5 and 1. It turns out, however, that if the shape of such music gets too disordered – in other words, the value of *a* gets too close to 0.5 and hence the shape looks too jagged – it begins to sound too much like a random arrangement of notes as opposed to interesting music. And this is precisely the point: the music we find interesting is neither too ordered – for example, like the simple patterns in *Three Blind Mice* – nor too disordered, such as we would get with a random-walk of notes moving up or down according to the flip of a coin. Exactly where we would each draw the line is a matter of taste, or dare I say sophistication. But the fact is that we all like our music to be not too ordered and not too disordered. Indeed, maybe we should actually classify our tastes by our preferred "*a*" value, or equivalently the corresponding fractal dimension *D*. It certainly might make choosing CDs for someone's birthday that bit easier.

In addition to the melody itself, fractals in music can arise due to the set of chord extensions used – such as a minor 6th – and also from the chord sequences and associated percussive rhythms. So even if the melody of some music might appear relatively simple, as long as the chord structure or rhythm section are sufficiently fractal then the end result can be just as interesting.

I just want to briefly come back to modern jazz before moving on to the major point of this section. Modern jazz is not to everybody's taste. But of all the different styles of music, I would argue that modern jazz is the closest to a true Complex System and hence closest to the musical equivalent of Complexity. Thinking back to the key components discussed in chapter 1, this makes sense. Modern jazz involves a spontaneous interaction of a collection of objects (i.e. musicians). It exhibits surprising emergent phenomena in that it is improvised, and hence what emerges in a given solo is a product of the actual feedback which that soloist receives at that moment in time. It is also an open system in that its best performances arise in an environment with audience feedback. It even has the feature of extreme behavior when, for example, the whole ensemble begins to mimic the pattern being played by a particular soloist, and hence there is a crowd effect in which the whole group begins to synchronize its phrasing. Above all, it has no "invisible hand" such as an orchestral conductor or an existing piece of melody that all the players are simply repeating.

Instead solos are built on the patterns, motifs or "licks" that a given player has in his memory, and which are then interwoven with original ideas in a truly complex way – all set against a loose backdrop of chord sequences. Just take a listen to the Brecker Brothers album *Heavy Metal Bebop*, in particular the track *Some Skunk Funk*. To me it embodies everything that Complexity is – and if you ever happen to see a transcription of Michael Brecker's tenor saxophone solos for that record, you will see fractals dancing before your eyes. If, however, you prefer something a little bit easier on the ear and more popular, listen to the patterns being woven in the alto saxophone solo played by Phil Woods on Billy Joel's *Just The Way You Are*.

The fact that we like fractal music is very interesting in itself, since it turns out that many things around us in Nature are also fractals in either space or time. Even our own heartbeat which is supposed to be so regular, is actually a fractal. Indeed the healthier your heart is, the more fractal it is since the added Complexity – over and above what would correspond to a strictly clockwork heart – allows the body to adapt to a wider range of possible situations. So our heart is a Complex System which produces a fractal output time-series – in other words, a fractal in time – and the fact that it lies between complete order and disorder gives it adaptive power. In fact, the notion that there are benefits to sitting on the edge between order and disorder, and hence having adaptive power, rings true in our everyday lives. As we all know, a workable daily schedule cannot be made too rigid or ordered since any small blip would then make the whole thing collapse. Instead, a certain level of flexibility is very important if we are to adapt to unexpected events. Our hearts follow the same tactic – and so does the music we find most interesting.

As hinted at earlier, there are also many everyday examples of fractals which appear as patterns in space as opposed to patterns in time. For example, the skyline of many cities and the typical shape of many mountain ranges are essentially fractals in space. Indeed the reason why so many spatial fractals appear can again be understood in terms of Nature and Society working in such a way that the resulting forms are neither too ordered nor too disordered. In mountain ranges, it is likely that whatever physical process caused the roughness of a particular mountain, also gave rise to a similar roughness on a nearby peak. Hence there is a persistence in the amount of order as we move from peak to peak, exactly as we needed for our walker in order to produce a fractal with a value of a between 0.5 and 1. Such persistence can also be referred to as a correlation in the behavior of neighboring parts of the same mountain range, just as the persistence in an output time-series can be referred to as a correlation in the output from one moment to another. In this language, what stops the melody of *Three Blind Mice* from being interesting is that it is just too correlated. A similar argument holds for the skyline of cities. It is likely that the height or shape of a given building would somehow have been chosen to fit in with those nearby – at least if the city planners have done their job. And since we seem to find fractals so aesthetically pleasing, judging from our tastes in music, it is little surprise that a pleasing city skyline will also correspond to such a fractal. Even art is a form of fractal, and the art that we tend to find most interesting is also in that middle-ground between complete order and complete disorder – in other words, between being completely boring and completely unintelligible.

So fractal shapes are a fairly ubiquitous feature of Complex Systems. In other words, they represent a fairly common emergent phenomenon from Complex Systems, in which the system behaves in time as though it is bouncing around in the middle-ground between complete order and disorder. In the same way, the shapes that a Complex System exhibits in space also appear to live in this middle-ground. However, fractals are not the only things that can arise in a Complex System. As we saw earlier, periodic behavior can also occur – so can purely static behavior. What makes a Complex System truly complex is the way in which it gives rise to such effects *and* how it manages to move between such different types of behavior. So while it is sometimes stated in the popular press that our lives are "on the edge of Chaos" and that everything in our world can be explained using fractals, such claims represent a gross over-generalization.

3.6 What happens when I'm not looking?

In the rest of the book, we will focus on real-world Complex Systems rather than the toy versions that we have discussed in this chapter. But before we do, it is worth thinking about how we actually observe a Complex System. After all, this is a very important issue given the standard approach of scientific research, which is to observe something, then measure it, then develop a model or theory, and then test this against the observations. Indeed if we are not observing the system properly, or if the way in which we are observing the system is unintentionally biasing our measurements and hence deductions, we will be in trouble.

The natural assumption to make is that the act of observing a Complex System does not in itself affect our interpretation of what we think is going on. But this can be misleading. Just think of the following two extreme examples. Suppose we want to investigate whether there is any pattern in the arrival times of trains at a particular station. Let's suppose that the whole rail system always runs perfectly on time but that we unwittingly turn up at a different time each day and do not know the precise time. Based on our observations, we would probably end up wrongly concluding that the trains themselves arrive at random times. As a second example, imagine something that is observed regularly and in the moment of being observed always has the same form. For example, imagine a child who is always in bed asleep between 8 P.M. and 6 A.M., and yet spends his days running around in an essentially random way. A visitor from another galaxy who only visited between the hours of 8 P.M. and 6 A.M., and who knew nothing about human life, might wrongly conclude that humans spent their lives in a very ordered fashion, asleep in bed. The moral of these two stories is that we have to be careful how we are watching. We must be careful not to add any Complexity which isn't already there, nor must we miss the Complexity which is actually there. This is an important practical point for Complexity Science as a whole, and has particular relevance for some of the applications discussed later in this book.

Chapter 4

Mob mentality

4.1 I'm a person, not a particle

The things that we observe around us – such as traffic, financial markets, and even ourselves – tend to occupy the middle-ground between order and disorder, making occasional forays toward one or the other and back again without the help of any "invisible hand" or central controller. It is the emergence of such properties that make a Complex System complex, and sum up what we mean by Complexity. The magic ingredient underlying these remarkable properties is feedback.

Memory is a form of feedback. It represents a feedback of information from an earlier point in time. Other forms of feedback include information being fed back from somewhere else in space – like a cell phone call from a knowledgeable friend. In our office-filing scenario in chapters 2 and 3, we saw that feedback could be introduced into the filing system via a single object – an intern. However, a Complex System contains a collection of objects with each object potentially giving and/or receiving feedback in one form or another at any given moment. Moreover, the feedback in a real-world system comes both from within the system itself as well as from the outside – and the net effect will therefore tend to be much more complicated than in the case of the single intern and the files. Hence the resulting behavior or dynamics of the system – as observed from the output time-series – will tend to be far richer.

We know that a collection of inanimate objects such as a pile of files – or the pile of socks in your laundry basket – cannot re-order itself without adding some feedback. In other words, it requires the helping hand of the intern – or some blood, sweat and tears on behalf of the owner of the socks. By contrast, collections of people, animals and even cells are able to do this, forming well-defined groups or crowds for example. This is because the individual objects in question are alive, and hence each is able to put in effort and energy to make decisions and take actions. In addition they each have some kind of memory of the past which can guide their actions. It is this intrinsic feedback generated by these objects, both individually and as a whole, which is ultimately the source of the Complexity which arises in collections of all living objects – from collections of cells which produce organs, to collections of organs which produce humans and animals, to collections of humans and animals which produce crowds.

But is there anything more that we can say about the behavior of collections of living objects such as humans? After all, people are complicated – yet somehow their combined decisions and actions give rise to well-defined effects such as financial market crashes and traffic jams. It sounds like a hopeless task. That is why sociologists, economists, anthropologists, historians, and political scientists are all so gainfully employed – often analyzing the same human phenomena over and over again, yet managing to provide a multitude of different explanations. The idea of complementing this process, or possibly even replacing it in some cases, by a scientific theory of collective human behavior sounds ridiculous. However, there are signs on the horizon that the task is not actually as crazy as it first sounds – for the following reason. Even though we humans are complicated in terms of our tastes, thoughts, beliefs and actions, the *ways* in which we are each complicated as individuals may not be so important when we are put together as a group. So even though there are many differences between all our different personality types, these differences may cancel out to some extent when we are in a large enough group – and hence the group as a whole behaves in such a way that these individual differences don't matter so much. This in turn opens up the possibility of building realistic simulations of collections of humans using a computer program containing populations of "software people". Such simulations may then mimic the overall behavior of a human Complex System such as a financial market or traffic – at least in terms of its general behavior. Indeed this is exactly what is now being explored in research laboratories such as ours across the world, exploiting the recent advances in virtual-world building as developed by computer game programmers, and by the so-called many-body mathematical machinery of physicists.

So although it might take volumes of books and TV documentaries to try to explain the complex life of someone like Winston Churchill, a randomly chosen collection of such famous people would likely behave in a similar way to a randomly chosen collection of the rest of us. A good example is given by *Big Brother* and *Celebrity Big Brother*: it makes little difference whether the individuals in the group are well-known celebrities or not, nor whether they are cooks, construction workers or clinicians – as a collection of humans, they seem to produce fairly similar dynamics when faced with the same everyday problems, as viewers are beginning to realize after putting up with series after series of these reality shows. In other words, a randomly chosen group of people tends to exhibit somewhat similar traits to any other randomly chosen group of people.

I am not saying that groups of people behave in a *simple* way. Nor am I saying that the ways in which groups behave is simply some scaled-up version of how individuals behave. Far from it – after all, the behavior of emergent phenomena such as traffic jams and financial market crashes does not typically reflect the behavior of any particular individual. What I am trying to suggest is that the overall behavior of such groups can be quite similar. In particular, even though the traits of any two individuals can differ wildly, the groups to which they belong can behave in quite a similar way. For this reason, traffic jams tend to look the same in Japan, U.K., U.S.A. and Australia, and so do financial market crashes – and yet the individuals involved can be very different. In other words, the ways in which collections of humans tend to "do" financial markets and traffic – and even, as we shall see in Chapter 9, wars and conflict – can be remarkably similar, despite their individual differences in terms of geographic location, background, language and culture. This is one of the reasons why the patterns which emerge from such Complex Systems can be so similar – or in techno-speak, *the emergent phenomena have some universal properties*. You can see how this might arise as follows: we are all sufficiently different that any particular collection of us is likely to include people with a somewhat opposite character to us. For real-world examples, just think of your own extended family, school or college class, or office colleagues. Hence in terms of the behavior of the collection as a whole, our individual quirks tend to get cancelled out to some extent. So the variation between groups of us is less noticeable than the variation between any two of us as individuals.

There is a very important consequence of this for researchers attempting to build virtual worlds containing populations of software people, in that these software people don't have to be individually as quirky as real humans. In short, they can be individually far simpler than real humans and yet still create a realistic overall collective behavior. And given that nobody knows how to create a real software person, this "safety in numbers" represents a fantastic simplification. Indeed, not only should this approach apply to collections of humans, but it should also apply to any other collection of objects which are individually complicated and hence exhibit a diversity of possible behaviors – for example, we should be able to apply it to cells in the modelling of a cancer tumor. This is the key idea that we pursue in the rest of the book, based on the extensive body of research which is now emerging in this area.

Our challenge in the rest of this chapter is to develop a generic, yet realistic, description of a collection of decision-taking objects such as people. Everyday human examples include whether or not to take a particular route home, whether or not to attempt going to a potentially crowded bar, whether or not to go to a particular supermarket, and whether or not to buy a particular stock. In fact such "whether or not" questions are part of a general set of decisions that we, and everyone else, are taking all the time – either consciously or unconsciously. Indeed, although many decisions we have to take are very complicated in their details, they can pretty much always be broken down into "whether or not" type questions. Put more formally, they can be broken down into "choose 0 or choose 1" questions. So, the two possible routes home can be represented by 0 or 1; selling or buying in a market can be represented by 0 or 1; not going or going to a potentially crowded bar can be represented by 0 or 1. And since 0 and 1 are binary numbers, we can refer to this type of "whether or not" question as a binary decision problem. Such binary decision problems are like games which we are all playing all the time. In these games, there is an underlying competition for some kind of limited resource, or food. It might be space on a road or in a potentially crowded bar – or in a financial market setting, it might be competition for a good price. It doesn't matter about the context: we each need to make a decision and hence take an action, and the net result of all our actions will generally determine which decisions were individually the best. But because of the limited resources, we cannot typically all be winners all the time.

Why is the competition for limited resources so important in producing Complexity in real-world systems? The answer is simple. In real-world situations where there is no competition, it matters little what decisions people actually make. In other words, if there is an over-supply of desirable resources – space on roads or in crowded bars or in supermarkets, money, fame, jobs, land, political or social power, food etc. – then it doesn't matter very much what we decide to do since we will still have enough of everything we need, and more. In such situations, we could each go round acting in whatever way we wanted, either cleverly or stupidly, and yet still end up with an embarrassment of riches. Hence there is no need to learn from the past, or adapt. The need for feedback then becomes pretty meaningless since we are all getting what we want all the time. The end result is that the collection of objects in question will behave in a fairly simple way. In particular, the lack of dependence on any feedback or interactions between the objects will make the overall system *non*-complex.

Most real-world systems of interest do feature some kind of limited resource – and the individual objects in the system will typically fight tooth and nail to get hold of it. This competition might even lead to local cooperation within groups in order to gain some kind of competitive edge – but the overall atmosphere is still one of competition. Examples include a financial market where traders are trying to compete for the best price; traffic where commuters are all trying to use a particular road; the Internet where surfers are all trying to access particular downloads more quickly; wars and terrorism where the different armed actors are all fighting for control of land or political power in a particular country. And remarkable things can then happen – for example, the emergence of crowds and opposing crowds which we call "anticrowds". We'll look at these more closely in a moment – but suffice to say, the spontaneous formation of such self-organized, collective phenomena is a true signature of Complexity. So not only do such binary decision problems represent a common everyday situation for all of us, it also turns out that they provide the perfect example of real-world Complexity. Furthermore, they provide the research community with an extremely challenging scientific problem which has, so far, no exact mathematical theory to accompany it. And yet all this richness arises from a situation as familiar as the daily commute to and from work, and a situation as fun as a trip to your favourite bar. So let's take a closer look at such binary decision problems, casting the discussion in terms of that all-important issue of a Friday night out at your favourite bar.

4.2 Thank God it's Friday

It is Friday night and there is a great band playing at your favourite bar or club. Actually they play there every Friday, and you like to go there as many Fridays as possible *provided* that you can get a seat. But how do you know in advance whether you will get a seat? The answer is, you don't. It is a small bar, and there is only limited seating – and there are also likely to be many others wishing to go as well. So what should you do? Make the effort to go all the way to the bar, only to find you can't get a seat? Or stay at home and risk losing a great night out?

Let's take a closer, more detailed look at this problem since it embodies what Complexity is all about. Indeed this is why Brian Arthur of the Santa Fe Institute first introduced it. Suppose that the bar gets very uncomfortable to be in if there are more than sixty people present – in other words, the comfort limit of the bar is sixty. Let us write that in mathematical form by saying that the comfort limit of the bar is represented by the symbol L, and therefore $L = 60$. How many other people are also making a decision as to whether to attend on a given Friday night? Again, we will represent this as a symbol – this time called N. Unfortunately for all the potential attendees, there is no way of knowing what that number N actually is. Probably it is OK to guess that more or less the same group of people want to attend each week, in other words N is the same number each week. But that still doesn't tell you what the number N actually is. For the purposes of our example, let's pretend that there are 100 people who want to attend on any given Friday night. In other words, or rather in our mathematical words, this means that $N = 100$. So there we have it: the comfort limit $L = 60$ and the number of people wishing to attend on a given Friday night is $N = 100$. And therein we have a typical everyday example of a collection of humans competing for some limited resource, which in this case is seating in a potentially overcrowded bar. We also have the archetypal Complexity problem in the making.

Now the fun and games begin, literally, because every one of these one hundred potential attendees is – whether they like it or not – playing a game: a gambling game, to be precise, since they each have to decide whether or not to go through all the effort of getting dressed up, arranging for a babysitter if necessary, going out to the car, driving to the bar, finding parking, walking up to the door of the bar . . . just to run the risk of being stuck in a bar which is over-crowded. At the same time, they could simply decide not to bother, and instead sit at home . . . just to run the risk of being told later that they missed a heck of a night out in a bar which had seats to spare. It is a gamble. Not only is it a gamble, but the correct decision as to what to do will depend on what everyone else decides to do. There is no absolute right and wrong decision – it depends on what everyone else thinks is right or wrong. If they all go to the bar, clearly the right decision is to stay at home. On the other hand, if everyone decides to stay at home, the right decision would have been to go to the bar. It is a competition for a limited resource, and not everyone can win.

In addition to its relevance to Complexity Science, it turns out that there is also an important consequence of this setup for standard economic theory. Specifically, there is no correct expectational model for this setup – if everybody made the same decision, it would automatically be the wrong decision since everyone would either stay away (in which case you should have turned up) or they all show up (in which case you should have stayed away). Hence the so-called "rational expectations" or "rational agent" model of standard economics breaks down. And since standard economic theory relies on the world being flooded with such typical people, our little example of bar-going turns out to be a huge wake-up call for Economics as a whole. Physicists don't get off much lighter either – any attempt at a so-called mean-field theory based on a similar notion of a typical bar-goer, will not work for similar reasons.

4.3 Knowing what it means to win

Let's develop this Friday night bar problem in a more scientific way. As we mentioned, this is an extremely important problem for Complexity Science, and as such is currently being studied by a large number of research groups from physics to finance, economics to sociology, computing to artificial intelligence, and geography to zoology. Irrespective of the specific application areas, all these researchers are interested in the same thing: the overall behavior of collections of decision-making objects, where the fact that there are only limited resources means that not everyone can win. Clearly an important ingredient is to sort out exactly what it means to win. In other words, how much reward, and similarly punishment, is there? Or put more simply, what is the "pain-gain" trade-off for each potential bar attendee?

It turns out that there are several ways of winning and losing in our bar problem. These are shown in figure 4.1. In particular, there are the following four potential outcomes each Friday night for every one of the $N = 100$ potential attendees:

First way of winning: You decide to go to the bar, and the bar turns out to be under-crowded because less than sixty other

Actions \ Outcomes	Over crowded	Under crowded
Go to bar	Lose	Win
Don't go to bar	Win	Lose

Figure 4.1 Knowing how to win, and knowing how to lose

people turn up. For example, suppose the total number of people attending on that given Friday is only fifty. Since fifty is less than sixty, this means that it was the right choice to have attended – you have won! We can represent this mathematically by saying that the total number of people attending on a given Friday t is given by $n[t]$, where in this example we have $n[t] = 50$ with a comfort limit $L = 60$. You therefore win in situations where $n[t]$ is less than, or equal to, L.

First way of losing: You decide to go to the bar, and the bar turns out to be over-crowded because more than sixty people turn up. In short, if you decide to attend in a given week t, you will lose if $n[t]$ is greater than L.

Second way of winning: You would also consider yourself a winner, if you decided *not* to attend and it turned out that the bar was over-crowded. In short, if you decide not to attend in a given week t, you will win if $n[t]$ is greater than L.

Second way of losing: You lose if you decide not to attend and it turned out that the bar was under-crowded. In short, if you decide not to attend in a given week t, you will lose if $n[t]$ is less than, or equal to, L.

Associated with each of these outcomes, we can imagine that there is some kind of payoff – not necessarily monetary, but in terms of satisfaction. Now of course, we all might assign different levels of satisfaction or dissatisfaction to this process. But since we can assume that all potential attendees are pretty much alike in terms of what they want, we can take winning as corresponding to gaining some kind of "unit of satisfaction" or "wealth", and losing as paying out a similar unit of satisfaction or wealth. In other words, it is just like those family games that we play at Christmas or on vacations – indeed, this whole process of deciding whether or not to attend the bar is just one big game.

4.4 To bar, or not to bar?

We now know what it means to win or lose in this game. But how should you go about deciding what to do? Let's start by thinking what would happen if you had no memory of what you had decided to do in previous weeks, nor do you have any information about whether the bar was overcrowded or not. In short, you have no feedback from the past. We are also assuming that you have no idea about the number of other people N who are interested in attending. Hence you'd probably be tempted to flip a coin to decide what to do this coming Friday. In fact that is the best thing you can do, in the absence of any other knowledge. Therefore you would be deciding based on the toss of a coin – in other words, randomly. Indeed, things wouldn't really change even if you knew some of the other possible attendees and were able to phone them to find out what they planned to do. It may turn out that they later change their minds. Furthermore, this is of course a competition for limited seating: so even if they told you what they planned on doing, should you automatically believe them?

Central to Complex Systems and hence Complexity is the presence of feedback. But so far, there isn't any – and this is completely unrealistic since we all carry round some kind of memory of previous events. Regardless of whether this memory is correct, partially correct, or totally misguided, it does exist and will tend to bias our decisions in a given situation. So as regular bar-goers, we will probably remember what we did on a few previous Fridays and we will probably also know what, in hindsight, would have been the best thing to have done on those nights. We will therefore each be keeping, either consciously or unconsciously, some kind of running tally of how many past successes we have had. Hence we will have some kind of notion of whether our methods of working out what to do – in other words, our strategies – need revising or not. The same thing would occur in a traffic setting, and in particular whether we choose a particular road home or not. We will remember whether we did or didn't "win" in a few previous commutes home, based on our own memories of what we did and based on what we subsequently heard from other people or from the TV regarding the traffic on those nights.

But how specifically should we add this feedback? There are many, many ways of doing this – but it turns out that there are two principal setups that have been followed in the research community and which are featured in the publications listed in the Appendix. It is best to think of these two setups as sort of "vanilla" versions, to which different bells and whistles can be added. In particular, the two setups exhibit differing amounts of systematic and careless behavior by the bar-goers – in other words, differing amounts of systematic versus random decision-making as with the intern in chapters 2 and 3. The first setup involves bar-goers who are not very systematic, but neither are they too careless. This case is discussed in depth in this section, since it corresponds to the middle-ground behavior which so many of us actually follow. The second setup involves bar-goers who are almost completely systematic and leads to a far greater frustration among the population (the notion of frustration was discussed in chapter 2). Despite the differences in the two setups, and regardless of the bells and whistles which one then adds, it turns out that the same general phenomena – in particular, crowds and anticrowds – emerge in both setups.

We'll focus here on the first setup involving bar-goers who are not very systematic in their decision-making, but neither are they too careless. In other words, they are not so systematic and reliable that they act like little computers – nor are they so random that they just use a simple coin-toss to decide their actions. Just as with the intern in chapter 3, we can mimic this middle-ground behavior with a coin whose outcome is biased by some feedback or memory from the past. However unlike our simple example in chapter 3, we will allow the specific way in which the coin-toss gets biased to differ for different people – and it will be through this individual bias that we will be able to mimic the real-world situation whereby a given population of people typically contains a diversity of character types.

Let's suppose that we are each carrying around a memory of our net number of past successes. In other words, we remember the tally given by the number of times we have won minus the number of times we have lost. If the number of losses exceeds the number of wins by too much, we will change the way in which we make our decisions – in other words, we will change our strategy. Just to be specific, let's imagine that "too much" is a number d. But what is our strategy? We will suppose that all the potential attendees know what happened on the previous m Friday nights – for example, this could be $m = 2$ corresponding to the last two Friday nights. More specifically, they know whether the correct action on the previous m Fridays had been to go to the bar or not. In mathematical terms, they therefore know whether $n[t]$ was less than or equal to L, or whether $n[t]$ was greater than L. Let's call these two outcomes 1 or 0, corresponding to whether the correct decision was to go to the bar or not. So an outcome 1 means that the correct decision was to have gone to the bar, while an outcome 0 means that the correct decision was to have not gone to the bar. In other words, we all effectively store 0's or 1's as a record of the past. For example, in the case where we remember the last two weeks' outcomes, our record of the past outcomes might be 11 which means that the correct decision was to have gone to the bar on each Friday night. The other possibilities are 00 which means that the correct decision was to have not gone to the bar on each Friday night; 10 which means that the correct decision was to have gone to the bar two weeks ago, but not last week; 01 which means that the correct decision was to have not gone to the bar two weeks ago, but to have gone last week.

Suppose that on a given Friday, the previous $m = 2$ correct decisions were 11. We will imagine that each of the potential attendees remembers what happened the last time that this particular pattern 11 occurred. It may have been only a month or so ago, or much longer – it doesn't matter. The same holds if the previous $m = 2$ correct decisions were 00, 10 or 01. In short, all potential attendees carry around a sort of crib-sheet containing a list of the winning decisions following the last time each pattern arose. This crib-sheet is therefore a source of common information.

So far, so good. Each bar-goer has a tally of past successes giving them a notion of how well their strategy is working – this represents their own piece of private information. They also have a crib-sheet telling them the correct decision following the last occurrence of each of the possible patterns that they will be faced with – this represents their public information. In fact, the public information could be broadcast by a public information system such as the radio or TV. All that matters is that all bar-goers know it, and it is the same for all of them.

Here is where their individual characters will now enter, by means of the biased coin which we mentioned earlier. Some of us might tend to think that given what happened the last time that a particular pattern of outcomes appeared, the same thing is likely to happen again – in other words, we believe that history repeats itself. On the other hand, others might think that it is precisely because it happened last time that the opposite will now happen. In other words they think that since the game is competitive, past history won't repeat itself and instead the opposite will happen. Many of us might also lie between these two extremes – but quite where we sit will vary for each of us, and we may as individuals decide to change where we sit as time goes by. We can mimic these various possibilities by assuming that each potential bar-goer has a coin which he tosses, where an outcome of heads means that he assumes that history will repeat itself. In other words, when faced with a given history such as 11, he simply looks at the crib-sheet to see what the correct decision turned out to be the last time that 11 occurred – and then he assumes that history will repeat itself if the outcome of his coin-toss is heads. By contrast, an outcome of tails means that he assumes that history will not repeat itself and instead the opposite will happen. In other words, when faced with a given history such as 11, he simply looks at the crib-sheet to see what the correct decision turned out to be the last time that 11 occurred – and then he makes the opposite decision.

To account for the different characters of the bar-goers, we can assume that the chances of throwing a heads or tails can differ for different people. In particular, we will say that the chance of throwing a heads, and hence assuming that history repeats itself, is given by a number p for a particular person. This number could be 60 percent for example, in which case that person believes that history repeats itself 60 percent of the time, or equivalently six out of every ten times. In coin-toss jargon, we would say that this is represented by a probability of $p = 0.6$ of throwing a heads. Likewise the chance of throwing a tails, and hence assuming that history does not repeat itself, is 40 percent. This corresponds to a probability of $(1- p) = 0.4$ of throwing a tails. So the p value that somebody has, acts like a kind of character description. Somebody with a value of $p = 1$ will assume that history always repeats itself. Somebody with a value of $p = 0$ will assume that history *never* repeats itself and that the opposite will always happen. Somebody with a value of $p = 0.5$ is literally sitting on the fence – they are so uncertain about whether history will repeat itself or not, that they are continually switching between the two opposing points of view.

Figure 4.2 shows what would happen in the case of only three bar-goers. In this example, history says that the correct decision will be to go to the bar. One bar-goer has $p = 0$ and hence always does the opposite of what history says, one has $p = 1$ and hence always does the same as what history says, and one has $p = 0.5$ and so always flips a regular coin to decide what to do. Depending on whether the outcome is heads, i.e. he goes to the bar, or tails, i.e. he doesn't go to the bar, the number of people in the bar will be 2 or 1.

Notice how similar this is to thinking about arrangements of files among shelves – which is precisely why we went through the discussion of inanimate or non-decision-taking objects first, before looking now at decision-taking objects. Moreover, this similarity means that every type of behavior that we have seen emerge in chapters 2 and 3 for collections of non-decision-taking objects, can also emerge in collections of decision-taking objects such as people – and more.

So what happens in this setup when we have lots of people? We can assume that everyone starts off with an initial value of p and that if they perform very badly over time, they will modify their p value in some way. Earlier we indicated how the effect of modifying p could be incorporated in a simple way – if a bar-goer finds

Figure 4.2 Just the three of us. The possible outcomes for a three-person population in which one person believes that history will always repeat itself (i.e. $p = 1$), one person believes that the opposite will always happen (i.e. $p = 0$), and one person continually flips between the two (i.e. $p = 0.5$). In this case, past history says that you should go to the bar. If the $p = 0.5$ agent's decision is to go to the bar, then there will be two people in the bar – hence the bar would be overcrowded if the comfort limit is 1. If the $p = 0.5$ agent's decision is not to go to the bar, then there will be one person in the bar – hence the bar will not be overcrowded if the comfort limit is 1.

that his net number of losses minus wins becomes equal to d, he changes his p value. His personal tally of losses and wins is then reset to zero, and he uses his new p value until it too needs modifying – and so on. As long as his net number of losses minus wins is less than d, he will continue to use his existing p value.

4.5 Crowds and Anticrowds

Since we are trying to mimic the general diversity of a population, it makes sense to start off our collection of people with a range of p values. One simple way of doing this is just to randomly assign each bar-goer with an initial p number between 0 and 1. Then we let the system evolve as described in the previous section. The big question is then this: what happens to this collection of bar-goers as time goes on? Given that no-one can win all the time, does the population gradually become so uncertain that they head toward $p = 0.5$? In other words, do people head toward completely random behavior?

Remarkably, it turns out that they do not. Pak Ming Hui of the Chinese University of Hong Kong has studied this problem in great detail with his own group of graduate students and in collaboration with my own group. It turns out that instead of heading toward random behavior, the bar-goers tend to head away from it toward the extremes. In other words, they become polarized into those who think that history will repeat itself, and those who think that the opposite will happen. The bar-goers who tend to shift opinions about whether to go with history or against it, tend to lose more and hence eventually change their p value. The ones with extreme values of p around 0 or 1 lose less and hence stay with those p values for a longer time. The net effect is therefore that the population of bar-goers tends to segregate itself spontaneously into two groups – one around $p = 0$ and one around $p = 1$. In other words, the population tends to split into a crowd that believes that history tends to repeat itself and hence sit around $p = 1$, and another crowd that believes that the opposite happens, and hence sit around $p = 0$. We call the former a *crowd* and the latter an *anticrowd* since they take the opposite action to the crowd. Figure 4.3 shows this effect in terms of a diagram.

But maybe you are thinking that all this is just a quirk resulting from the way in which we have set up the problem? Maybe things change if we change the value of m, i.e. the number of past outcomes featured in the crib-sheet of public information? Or maybe it matters how people change their p value? Interestingly, it turns out that these things don't matter – in other words, this surprising segregation of the population into a crowd and anticrowd is not a quirk. Instead, the emergence of the crowd and anticrowd turns out to be a very generic feature of such competitive games. In fact, it doesn't matter what the value of m is, the same shape emerges as in figure 4.3. And since this bar problem is just one example of such decision problems, the emergence of crowds and anticrowds will arise in traffic, markets and all sorts of other systems. In other

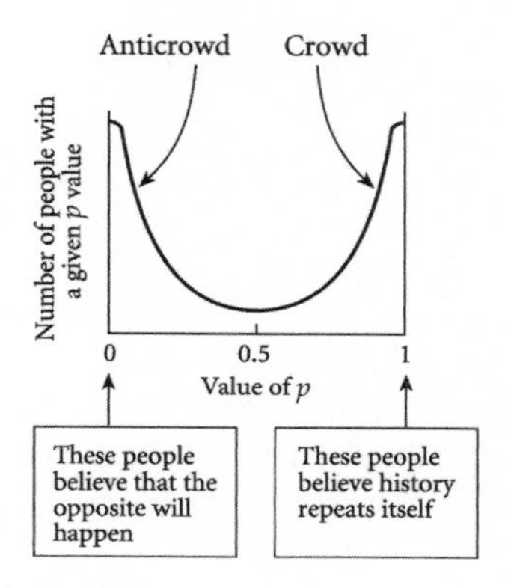

Figure 4.3 We are naturally divided. The final arrangement of a collection of people, in the case of a bar where the comfort limit is around half the number of potential attendees. This shows the emergent phenomenon of a **crowd** who think that history repeats itself, and an **anticrowd** who think that the opposite will happen. Hence the population polarizes itself into two opposing groups. This polarization of the population represents a universal emergent phenomenon. It will arise to a greater or lesser extent in **any** Complex System involving collections of decision-making objects such as people, which are competing for some form of limited resource.

words, it is a truly emergent phenomenon and is likely to be quite generic across different domains of application.

The crowd-anticrowd phenomenon has a very important consequence. The fact that people in the crowd around $p = 1$ will typically be taking the opposite action to the anticrowd around $p = 0$, means that as far as the behavior of the population as a whole is concerned, the actions of such opposite personality types tend to cancel each other out – which is exactly what we had earlier claimed would happen. This becomes particularly important in a financial market setting, for example, since it means that the number of people deciding to buy at a given moment, will essentially cancel out the number deciding to sell. To understand the consequence of such a cancellation, just think of the housing market – if there are an equal number of buyers and sellers, the price will not tend to move very much since supply and demand are balanced. There is no large excess of buyers tending to push the price up, nor is there a large excess of sellers tending to drive the price down. The resulting market will therefore tend to have small price-changes and hence a small volatility, or small fluctuations.

We have uncovered a truly remarkable effect. Even though there is no communication between the people in the game, and even though the game is competitive, the fact is that the system as a whole manages to self-organize itself in such a way that the fluctuations are smaller than for the case where everyone tosses a coin. More generally if we assume that fluctuations are bad for the system in some way, then we have shown that through competition the system as a whole does better than if the individual objects were to have behaved independently (i.e. by tossing a coin). The combination of competition for a limited resource and the fact that the people all receive the same sort of feedback, has given rise to a system which appears to be controlled by an "invisible hand" – but isn't. Looking back to the engineering applications mentioned in chapter 1, we can see why such self-organized control via competition and without the need for any communication or collaboration between the objects, would be so interesting for engineers. It offers them the possibility of mimicking the controlling effect of an "invisible hand" without the hassle of having to build or maintain a central control unit. In other words, they can remove potentially harmful or dangerous fluctuations in a system, simply by making the individual components competitive.

Even if we change the form of the public information being fed back to the agents – in other words, the crib-sheet concerning the past outcomes – the result is the same as long as the information that all the people receive is the same. In short, it doesn't matter whether the information is indeed past information or is instead someone's prediction, or a radio announcement, or a rumor, or even some false information. The fact that everyone receives the same information, and hence reacts to it using their respective p value, means that everyone is effectively joined together by this common information – regardless of where it came from, and regardless of whether it is right or wrong. Hence the feedback takes the form of common information, but it doesn't matter whether the actual information fed back is correct or not – it just matters that everyone receives the same information.

So far I haven't mentioned how the value of the bar's comfort limit L affects the results, or how the value of the net losses d might change things. Again it turns out that these don't have a great effect. As long as L isn't so close to the number of potential attendees N that the competition effectively goes away, then the same segregation into crowds and anticrowds will also arise. The only thing which will disrupt the formation of crowds and anticrowds is if we make the penalty for losing outweigh the reward for winning. This effectively changes the whole game since people will then tend to change their p value so rapidly that the population never settles down to a steady situation. By contrast, making the reward for winning outweigh the penalty for losing, does not disrupt the emergence of the crowd-anticrowd phenomenon.

4.6 Another frustrating experience

We just found that the population self-segregates by moving toward the extreme p regions. People who stick to the same type of decision in every turn therefore perform better than the uncertain people who choose randomly in each turn. We can understand this effect by again considering just three people. In fact we will go one step further and also only consider three p values: $p = 0$ corresponding to always going against the public information, $p = 1$ corresponding to always going with the public information, and $p = 0.5$ corresponding to someone being so cautious that they effectively flip an ordinary coin in order to decide. Let us suppose that for this problem of three bar-goers, the comfort limit is one. In other words, $N = 3$ and $L = 1$ which means that there are more than twice the number of potential attendees as there are seats in the bar – so this certainly embodies the notion of limited resources.

If we imagine the situation in which all three people have $p = 1$, then they all make the same decision and hence take the same action. Therefore all three people either go to the bar and it is overcrowded, hence they lose, or they all stay away and the bar is undercrowded, and hence they lose again. So this arrangement with them all having $p = 1$ is unlikely to last very long. Likewise, the arrangement with them all having $p = 0$ is equally unfavourable, and hence also unlikely to last very long. If on the other hand they all have $p = 0.5$, then there is at least a chance that one person will win by having two people throw a heads and one throw a tails, or vice versa. However there is also the possibility that they all throw a heads, or all throw a tails, and hence all take the same decision and hence lose once more.

A better arrangement, at least in terms of guaranteeing that not everybody loses, is to have two people with $p = 0$ and one with $p = 1$. That way there will always be one person who wins. Either two people attend the bar, and hence the one who doesn't will win – or only one person attends the bar and hence he wins. Similarly, the arrangement with one person having $p = 0$ and two with $p = 1$ also guarantees that there will always be one person who wins. These are the top two arrangements in figure 4.4. In addition, consider the arrangement with one agent having $p = 0$, one with $p = 0.5$ and one with $p = 1$. This is also shown in figure 4.4. Depending on the random outcome of the coin-toss of the $p = 0.5$ person, either the $p = 0$ or the $p = 1$ agent will lose. But again this guarantees that one person will win every turn. So the three arrangements in figure 4.4 are the ones in which one agent always wins. And if the system moves between these arrangements over time – very much like the system of files moved between arrangements in chapter 2 – the person winning won't always be the same. Instead, winning will be shared around the three of them. Hence it makes sense that the average distribution which is observed should be some kind of average of these three favorable arrangements, and hence favorable to the three-person population overall. We can see from figure 4.4 that an average of these three arrangements will indeed be biased toward the edges $p = 0$ or $p = 1$, rather than toward the center $p = 0.5$. The resulting U-shape at the bottom of figure 4.4 closely resembles the result seen in figure 4.3, thereby explaining the segregation phenomenon into crowds and anticrowds.

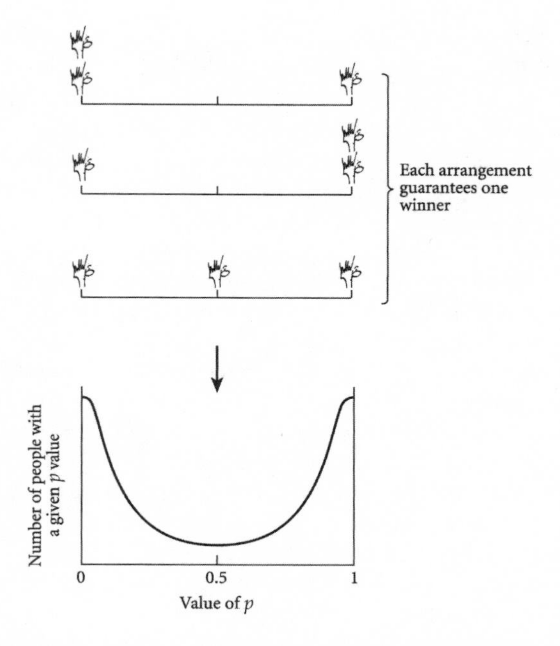

Figure 4.4 Three possible arrangements of three people among the three p values 0, 0.5 and 1. Since each of these arrangements will always result in two people making the same decision and one making the opposite decision, each arrangement will always produce one (and only one) winner for the situation of a bar with a comfort limit of one. Notice that the top two arrangements have nobody at $p = 0.5$, hence an average over these three arrangements will give the U-shape which is shown at the bottom. This then explains the U-shape observed in figure 4.3.

So we can now see why a collection of people will segregate themselves into a crowd and anticrowd. It is a truly emergent phenomenon and is characteristic of a collection of decision-making objects competing for some limited resource. Hence this idea of the emergence of crowds should arise across the entire range of real-world problems discussed in this chapter and in chapter 1, most notably within financial markets and traffic.

Now, if we had restricted the possible p values that a particular person could take up, we might have prevented certain of the arrangements in figure 4.4 from occurring, or at least reduced the probability that they would occur. This is exactly the effect of frustration that we saw in chapter 2. It is clear that if favorable arrangements are prevented, the system will perform less well overall in that the actual number of winners falls below the maximum possible number of winners. In other words the system will tend to under-utilize its resources, which is another signature of frustration. In such cases of frustration, the U-shape corresponding to figure 4.3 typically will be prevented from fully emerging. Instead it will be distorted in some way according to the precise nature of the frustration. In other words, either the crowd or anticrowd may dominate. Since the cancellation between the two will then be reduced, so the output – such as the price in a financial market setting – will move around much more. In other words the volatility or, equivalently, the fluctuations will be larger than before. The references in the Appendix include many studies of this crowd and anticrowd phenomenon in a variety of different situations, and result from an ongoing research program with Pak Ming Hui at the Chinese University of Hong Kong. In each of the different situations studied the results of the computer simulations can be understood and explained using the mathematical theory of crowds and anticrowds – the so-called *crowd-anticrowd theory*.

Finally, we will discuss very briefly the second popular set-up for adding feedback into such collections of decision-making objects. We mentioned in section 4.4 that this other set-up involved the bar-goers acting essentially systematically. Once again, the potential bar-goers all have access to the past m outcomes – but now they hold a couple of strategies, each of which is a fixed response to each of these outcomes. Because of this fixed response and the fact that people only have a few strategies to choose from – as opposed to the whole range of p values in the first set-up – the amount of frustration is large. In particular, the crowd tends to be much larger than the anticrowd. There is a regime of this set-up where this cancellation is reasonably large but it only occurs for a small range of m values. In addition to ourselves, this particular set-up has been studied in detail by D. Challet and Y.C. Zhang of the University of Friburg, R. Savit and R. Riolo of the University of Michigan, and D. Sherrington of Oxford University.

Given these set-ups, we have been able to explore a large number of variations and generalizations of such binary-decision games. These papers are listed in the latter part of the Appendix. For example, we have been able to explore whether someone who is fed larger packets of information will automatically find himself better off *at the expense of* others in the population who are being fed with less information. Interestingly the answer turns out to be no. The reason is that information becomes like food – and someone who feeds off of large chunks will tend to ignore the crumbs which are then left lying around for these others to eat. Hence they can all happily coexist in a sort of information ecology without getting in each others' way. Even more interesting is the finding that someone who is feeding off of such large chunks may unwittingly be generating smaller crumbs themselves – very much like someone who is eating large chunks of food will tend to leave small crumbs lying around. So there is a very real sense in which everyone can benefit from diversity within the population.

4.7 Evolution management: Engineering the future

So now we have understood how Complexity can emerge in collections of decision-making objects such as people. But what about our goal of actually managing such a Complex System so that we avoid any unwanted behavior? It turns out that the prospects are good – in particular David Smith of Oxford University has shown, using sophisticated mathematical analysis, that this is indeed possible. The details are in his research paper listed in the Appendix, but without going through all the steps I will give a sense here of what he has done.

Imagine a car with shuddering wheels. Obviously this would be a potentially dangerous situation, but we all know how such a car could be driven in practice: grip the steering wheel firmly to help reduce the shudder, and then turn it right or left as usual. As long as you are gripping the wheel firmly and hence are able to make the wheel turn more than it shudders, you will make the car go right or left as desired. Now imagine that this car has an underlying "Complex Systems" design; as a consequence, you might neither have complete knowledge of, nor direct access to, the steering mechanism itself. Instead, let's suppose that there is a very complicated set of inter-connected levers between you and the car's wheels – in other words, in true Complex Systems-style, the steering mechanism consists of a very complicated arrangement of interacting objects. So now you have two problems. First, there is the prediction problem: you need to predict where the car is going in the next few seconds in order to determine whether it is heading toward danger or not. Second, there is a control, or management, problem whereby even if you manage to work out where the car is going to move, and hence whether to try to steer it, you are still left with the problem of what levers to adjust in order to produce the desired steering effect.

Of course, if this were a standard car you would probably try instead to find a mechanic who has knowledge of, and can gain access to, the relevant parts of the vehicle. Given all the prior available information about the make and year of car, he would then know how to fix the problem – even if this meant the maximally intrusive step of taking the car off the road, stripping it down to its component pieces and rebuilding it. But such levels of intervention are not generally possible in real-world Complex Systems, nor indeed is the precise nature of their constituent parts and interactions generally known. Hence, even if one could work out what intervention is actually needed, its implementation could only ever be indirect, infrequent, and generally broad-brush in terms of its level of specificity. And to make matters worse, most real-world Complex Systems cannot be "shut down" – hence any such interventions would need to be carried out in real-time, like a car being repaired while it is speeding along a highway.

Despite these seemingly insurmountable challenges, David Smith's research shows that such online Complex Systems management is indeed possible. In particular, he has shown that both the prediction problem and the subsequent control/management problem have relatively straightforward solutions for the type of Complex Systems considered earlier in this chapter based on multi-object competitive games.

In terms of the prediction problem, David has shown that with relatively little knowledge of the past behavior (i.e. the global output), one can produce corridors into the future along which the system is then very likely to move. These corridors have two important features: their width, which he calls the Characteristic Stochasticity, and their average direction, which he calls the Characteristic Direction. In most real-world Complex Systems, from biology through to financial markets, the observed dynamics is so complicated – more specifically, the way in which the system moves between order and disorder is so complicated – that neither the width nor the average direction of these corridors will be constant. They will change in time and that is what makes traditional time-series prediction schemes so poor at predicting the future of real-world Complex Systems. However, David's work shows that with relatively little information about the past global output – in particular, without knowing precisely what the individual objects are each doing – he can produce such corridors into the future. The system's subsequent behavior is such that it then moves along these corridors with remarkable accuracy.

David's work shows that as time evolves, the widths and average direction of these corridors can vary significantly. In particular, there are moments when the width is much larger than the average direction, and vice versa. This just reflects the feature that I have emphasized in this book, that a Complex System exhibits pockets of order. In particular, it continually slips between order and disorder and hence so does the predictability of its future movements. In a situation where the width of the corridors is larger than the average direction, it would not be wise to make a firm prediction as to a particular direction for future movements – as for example, in trying to predict whether a price goes up or down in a financial market. Having said this, even a small "edge" may be enough in a financial setting: you can still make money even if you can't predict the direction of future price movements all the time. I will return to this in chapter 6 when we discuss financial markets. But for now, we can just think of David's corridors as providing an accurate view of how risk in the system evolves as the system itself evolves. In short, the behavior of these corridors over time indicates the appearance and disappearance of pockets of predictability, which in turn reflects the emergence of pockets of order in the system.

Now comes the control/management problem. As suggested earlier, just because we might know our system is heading "off course" it doesn't necessarily mean we can do anything about it. One might suspect that we have to master all the parts of the system in some way in order to make a significant change – and this sounds like an extremely intrusive level of control. Again, David's work shows that such extremes are not always necessary. He finds that it is typically enough to make a few general "tweaks" to the population's composition. In particular, we don't have to know very much about the precise composition of the population nor the precise change that we are making. The secret lies not so much in how big the tweak is but *when* we make it. In particular, we do not even need to have direct access to all the objects in our system – after all, they may be largely hidden such as cells in the body, robots searching a dangerous area for bombs or mines, or collections of spacecraft sent into outer space. All we generally need to do is tweak the subgroup that we do have access to. Even in a scenario where we don't have access to any of the objects, David has shown that we can then simply add some more objects to the collection in order to induce the effect of steering the system.

Setting aside "Complex System" cars, it turns out that there are many potential applications for such online systems management. Next generation aircraft will have so many interacting parts that the pilot will have no hope of assimilating all the available information. In short, the plane will be "out of control" since no single human or computer will be able to respond quick enough if things go wrong. So how should this system be controlled? One approach is to use a collection of competing miniflaps located along the wing, and then manage them in real-time according to David's scheme. Indeed this is exactly the approach being pursued by Ilan Kroo at Stanford University. In human biological systems, doctors increasingly find themselves having to manage complicated conditions to do with overall levels of physiological activity in the body. Worse still, the precise nature and level of such activity may be unknown to the doctor at any one time. For example, the precise level of immunity in the immune system cannot easily be measured, nor can the level of heart activity or mental behavior. This in turn provides a possible connection to so-called dynamic diseases such as epilepsy and their online management. Likewise, a seizure involves a sudden change in the activity of millions of neurons. Feedback control of seizures would require an implantable device that could predict seizure occurrence and then deliver a stimulus to abort it. This sounds a very intrusive procedure. But David's work raises hopes that it might be possible to develop a "brain defibrillator" which delivers brief but effective electrical stimuli over a small part of the brain, rather than intrusive control over each and every one of the constituent "agents" (i.e. neurons).

Another possible biological application is the treatment of cancer tumors. Inside each tumor there is an ongoing competition among cancerous and normal cells for two limited resources: nutrients in the blood supply and space to grow. A fully invasive procedure to remove a tumor might be so disruptive that it would actually promote mutation – for example, cutting out only ninety percent of a tumor might actually be even worse than leaving it untouched. By understanding how the overall population behaves one could apply David Smith's "population engineering" to a small group of the cells. Without being one hundred percent accurate – which is anyway impossible to do – David's work suggests that one might then conceivably be able to steer the tumor toward safer territory. Even in the immune system, where the body self-regulates itself as a result of the interaction of hundreds of different biological objects, one might be able to tweak one part of the system so that it modifies the overall behavior. For example, it might be possible to control autoimmune diseases such as arthritis, where the body attacks itself, by adding a vaccination against something entirely different. Given that all these entities or objects are interconnected, then by biasing the immune system population the overall system could be steered off in a given desired direction. Such minor and indirect intervention could prove to be enough if it is carried out at the right place and the right time.

There are interesting applications in the area of finance. First, an institution such as the Bank of England or the U.S. Federal

Reserve – who are not out to predict financial market movements *per se* – could, if necessary, step in to apply a small amount of influence to a small section of the trading population. They could thereby steer the system – in particular, the market index, or exchange rate – out of trouble. This is of course known to be possible if the intrusion is allowed to be massive – after all, if one makes trading illegal then the market will stop moving since there are now no trades being made. However, the implication of David's work is that this can be achieved without over-intrusive intervention – and hence at relatively low cost. Second, suppose that there is a fund manager holding lots of stock. These stock effectively compete with each other for their share of the fund's portfolio. If the fund manager sees that the value of the overall portfolio is momentarily heading downward, she could of course sell all the different stock and buy new ones. However, this is very expensive (i.e. very intrusive) because of transaction costs. David's work shows that she might be able to tweak this portfolio in real-time, by buying/selling a very small amount of stock in order to steer it clear of danger.

Next-generation "smart" technologies can also benefit. Consider a population of autonomous agents competing for a limited resource, as discussed in earlier chapters – examples of which include a cluster of extra-terrestrial craft, a cluster of anti-explosives robots checking out a building, or even a cluster of nanobots within the human body. Reprogramming these agents might be impractical or impossible and, as such, some other form of control would be necessary. David's work shows that this control could come in the form of a "vaccination" by injecting additional agents into the system. If the composition of this vaccination is chosen wisely, the competitive nature of the agents is such that their subsequent interactions with the rest of the population generate an overall steering effect via feedback. This approach could also benefit systems where the sheer number of participants renders conventional control impractical. Thinking further into future technology such control philosophies might even be appropriate for whole surfaces of materials which are coated with interacting active agents – so-called "smart surfaces" – or even new designs of "smart matter".

It may even help out with understanding and eventually controlling global warming. Typically there is no feedback between the actions of humans and the weather. The fact that many people will go and lie in the sun does not affect the chances of it being sunny. However, the actions of our society as a whole do indeed seem to be changing the weather over the longer-term. The reason is that human consumption produces waste products such as greenhouse gases which rise into the atmosphere and might eventually affect global weather conditions. The weather itself results from a complicated interaction between air and water – or more specifically, the changing temperatures of the oceans, air and land masses – so it is likely that this will be altered over the longer-term by our collective actions. Global warming. David's work suggests that we might "undo" these effects, or at least moderate them, without full control of the climatic system or knowledge of what its individual components are doing. For example, an upcoming flood, hurricane or drought might be weakened by some form of atmospheric intervention – possibly by injecting a harmless gas of particles into the air to either promote some pre-emptive rainfall or alter current cloud formations.

But for the final word on this let's turn to Hollywood. In particular, you might have already picked up on a similarity between this research and the movie *Minority Report*. The movie's storyline features a collection of "precogs" – which is shorthand for precognitives – who make fairly blurry predictions about the future. While the movie only has three precogs, it turns out that David Smith's mathematical model for constructing future corridors is analogous to a population of many such precogs. When this population of precogs don't agree – which in the movie was the origin of the secretive "Minority Report" – the corridors are very wide and have no definite direction. Hence the future becomes unpredictable. By contrast, when the population of precogs do agree – which in the movie was officially always the case – the corridors are narrow and have a definite direction. Hence the future becomes predictable.

Chapter 5

Getting connected

5.1 Knowing me, knowing you

We have just seen a collection of competitive, decision-making objects, such as people, self-organize themselves into crowds without the need for any "invisible hand" or central controller – as if by magic. What made this even more amazing was that it was entirely unintentional. Nobody should want to be part of a crowd if they are competing with each other. Yet crowds emerged – and they did so because everybody was being fed the same global information and they were all competing for the same limited resource, such as space on a particular road or a favorable price in a particular financial market.

This makes sense for Complex Systems like financial markets and traffic where people don't generally know each other and also don't know how to contact each other. But people tend to be social animals – and in many examples of human Complex Systems, individuals may well try to make private contacts and form alliances or coalitions of some kind. In other words, they begin to interact directly with some other sub-group. In short, they form a *network*. And the same holds for the animal and insect kingdoms.

The added feature of local contact and communication, and hence interaction between individuals, has led to the topic of networks becoming important in the study of Complexity. A network tells us who is connected to who, and therefore who is interacting with who and what their interactions are. Such a network may also play a role in feeding back information from one part of the population to another – for example, cell phone calls can instantly connect people who are geographically very far apart. In chapter 4 we talked about feedback which comes in the form of common, or public, information – typically from the past. Here we are opening up the possibility that feedback may also come from different points in space – moreover, that different people may have different types and/or different amounts of feedback depending on who they are connected to in their social network.

Scientists have so far focused on static networks. In other words, they have focused on looking at the entire set of connections that have appeared over some particular span of time instead of focusing on when these connections actually appeared and disappeared. But there is a big problem with this approach. Lumping together connections means that you lose any information about the order in which connections appear and disappear. And yet, this information about *when* things happen is very important.

Just imagine a collection of people among whom a rumor is being passed – or worse, some kind of virus is spreading. Let's focus on three people, and let's suppose that there is a particularly nasty virus going around. Imagine that persons A and B are not yet infected, but person C is. If you are person A, it really matters whether person B spends the evening with person C before or after spending the evening with you. If it is after, then you are safe – but if it is before, watch out.

It is for this reason – especially for the spread of something such as information, or a rumor, or a virus – that the network that you sit in is so important. Above all, it really matters how and when you are connected to who. It is also for this reason that the network research field has become incredibly active recently – mainly with the focus on large networks involving many objects. Particular pioneers in this field include Mark Newman of the University of Michigan, Duncan Watts of Columbia University, Steve Strogatz of Cornell University, and Albert-Laszlo Barabasi of the University of Notre Dame. These peoples' research papers are available at http://xxx.lanl.gov. Here I will only refer to specific results which are necessary for our Complexity story.

5.2 Small worlds, large worlds and in-between worlds

A network consists of a set of nodes, such as the three people described earlier. Depending on the network being studied, some or all of these nodes may be connected together by links. A network therefore gives us a visual picture of how a collection of objects are either connected or interact. Based on this reasoning, it follows that many of the things around us in our everyday lives represent examples of networks, from transportation networks and information networks through to social networks – and even voting networks. (See for example, the downloadable research paper listed in the Appendix which uses network analysis to uncover voting biases and cliques in the contest that Europeans love to hate – the Eurovision Song Contest).

Because of the wide range of possible internal interactions and behaviors that it can exhibit, a Complex System may produce a wide range of network shapes or "structures". In particular types of Complex System, the ways in which the network connections are arranged may seem to follow a particular pattern, for example, many social, transport and information networks have particular objects (i.e. nodes) with many more connections than others – in fact, an abnormally large number. These objects act as hubs to which many other objects are then connected. For a concrete example, just think of an airline such as *Continental Airlines* which has hubs in Newark and Houston. You may even be stuck in one as we speak, though I very much hope not. In addition, we probably all know someone who has so many friends that she needs a PDA to keep all their contact details – and we also know of others that have so few friends that they can remember their contact numbers by heart.

An important scientific question concerns the extent to which Nature does or doesn't favour centralized network structures. We can think of this in our everyday lives in terms of road planners choosing to either place a ring road around a city, or to add additional roads through the center. It turns out that biological systems such as fungi have to solve such a problem all the time. A fungus is essentially one big living network with no central brain or stomach. As such, it has to supply food to all parts of its network all the time – very much like a major supermarket chain has to continually resupply its stores with food products. It is therefore interesting to see exactly how biology, and in particular the fungus, manages to deal with such supply-chain problems.

The network structure phenomenon which has attracted most attention is arguably the so-called "small world" effect. It is mimicked in the earlier example I gave, using three people A, B and C. Even if person A does not know person C directly, the fact that A knows B and B knows C means that A and C are indirectly connected. Suppose that A and C happen to meet for the first time and in casual conversation they find out that they both know person B. They are then both likely to conclude that "It is a small world". Now imagine many people – in fact, think of the entire population of the world. We all tend to live a rather clustered existence: in different towns which are within different states or departments, which are in turn within different countries, and on different continents. Yet the shortest path between any two of us is still remarkably small on average. This was most famously demonstrated in 1967 by American psychologist Stanley Milgram. Milgram sent a number of letters to people in Nebraska and Kansas, asking them to forward the letters to a particular stockbroker in Boston. However, he didn't reveal the stockbroker's address. Instead, he asked the recipients to forward the letter to people who they knew, and who they suspected might be "closer" to the stockbroker as a result of their profession, location, or social circles. Milgram then let them get on with it, with the result that many of the letters arrived to the correct destination after very few forwardings. Six, on average. This is the origin of the term "six degrees of separation" and suggests that although objects might belong to quite different clusters (e.g. an office worker in Kansas, as opposed to a stockbroker in Boston) the average path length between them is typically very short. So although there are many people in the world, and the world is organized into clusters or communities, it is a small world in terms of who knows who.

A lot of current research effort is being spent on working out exactly how humans are connected. Given that Complexity features collections of interacting objects with feedback, the nature of such connections – and in particular what information they might carry – is clearly important. However we have to be very careful about how we define these network connections. In terms of friendship networks, for example, person A and person B may hate each other and therefore not be connected at all. However in terms of a network describing the transmission of a virus, they may well be connected if they happen to take the same bus by accident. After all, viruses don't require that the people that they pass between actually like each other. For this reason, we will be less concerned in this book with how things are connected and more concerned with the effect of such connections. Each of the networks which we discuss is equivalent to a collection of interacting objects which are competing in some way. Hence, in addition to being networks, they are also all examples of Complex Systems.

5.3 The importance of networking

We begin our peek at networks right down at Nature's smallest and by far weirdest level: the nanoscale level of quantum physics. It turns out that on this scale there is an extremely important network at work – from the leaf of spinach in your salad through to the entire Amazon Rainforest. It is the network of photosynthesis – or more specifically, the network of proteins inside each leaf which carry the energy of the light arriving from the Sun to a place within the leaf where it can be converted into food for the plant. However, even though photosynthesis is one of the first things that kids are told about in Science class, it continues to throw up many surprises for scientists. In particular, recent research suggests that Nature might actually be utilizing some clever network queue-management trick, in order to control the arrival rate of the light energy to the reaction centers where the food is produced.

More generally, biological Complex Systems show a wealth of network behavior. Just as the commercial world uses transport networks to distribute goods, or the Internet to distribute information, Nature uses networks to distribute the nutrients necessary for life – from the blood-flow through veins and arteries in our own bodies through to the nutrient-flow in forest fungi. Indeed, forest fungi represent a quite remarkable example of networks in that they consist entirely of a maze of tubes, and yet manage to stretch for miles over the forest floor – like some kind of natural "Wood Wide Web". But what is even more remarkable is the fact that biological networks such as fungi tend to reconfigure themselves over time in response to the supply of nutrients being transported. In other words, the system's food transport network affects its structural network and vice versa. Just imagine what would happen if we could get man-made networks to do this? We would have road networks which could rearrange themselves according to their traffic flows. This explains why scientists and engineers are so interested in understanding Nature's networks; in addition to a general fascination about how they operate there is the hope that we can pick up a few hints about clever network designs.

The tricky thing for researchers studying the Complexity of systems such as fungi is that there is no easy way of knowing which tubes in the fungus (i.e. roads) are being used to carry food (i.e. cars) at any given time. There is no such thing as being able to view the food flow from above using a helicopter, as they do for eye-in-the-sky traffic updates. All you see from above are closed tubes, and therefore you don't know what is flowing down what, and when. Fortunately researchers such as Mark Fricker at Oxford have managed to tag the food particles in the fungus so that they emit light – like a moving flashlight – which helps us understand where the food is flowing. But a big open question remains. How on earth do biological networks such as fungi manage to continually reroute food supplies without any centralized resource manager? This, after all, would be like Tesco or Wal-Mart never having to oversee the supply of goods to needy stores, but instead just sitting back and "letting it happen". Researchers also want to know to what extent biological systems use or avoid potentially congested hubs – and whether this knowledge could then be used to tackle congestion in man-made networks. Likewise, an understanding of the underlying nutrient supply network could help doctors in the diagnosis and treatment of potentially lethal cancer tumors, and in treating disorders such as an AVM

(Arterio-Venous Malformation) where the brain becomes starved of nutrients as a result of short cuts in the network of vessels.

On the level of groups of humans, a particularly important network is that concerning the transmission of viruses. At the time of writing this book, bird flu looks set to invade Western Europe – and scientists fear the possibility of it combining with a more typical human flu virus, thereby creating a superbug that could be transmitted easily between humans. Just as it is important to understand the actual biology of the virus, it is also important to understand how it spreads on a network.

Our society's safety is currently threatened by global networks of terrorism, crime and insurgency. It turns out that most modern conflicts represent a Complex System. Each is an evolving ecology with various armed insurgent groups, terrorists, paramilitaries and the army. In short, there are many interacting species which are continually taking decisions based on the previous actions of the others. In addition, these conflicts are being continually fed by an underlying supply network whose "nutrients" involve mercenaries, arms, money, drug-trafficking and kidnappings. Indeed, just as in a fungus and even a cancer tumor system, it is possible that these underlying nutrient supply chains have self-organized themselves into some reasonably robust structure, thereby making it even harder to remove or control them.

5.4 How things grow: Is it really all in the genes?

One of the most fundamental issues facing scientists is to work out the extent to which the macroscopic network structures observed in naturally occurring Complex Systems result from instructions within the genes at the microscopic cellular level. Crudely speaking, it is commonly assumed that genetics dictates structure and that structure then dictates function. But is it *really* all in our genes? This is something that David Smith, Chiu Fan Lee, Mark Fricker and Peter Darrah have been looking at very closely. They use the fungus as a particular example since they can inject tagged food into it, and thereby follow the flow of this food traffic through the fungal tubes or "roads".

They recently found something quite remarkable. Starting with a simple biological function – namely the passing around of food – they have shown mathematically that network structures emerge which closely resemble a wide class of multi-cellular organisms such as fungi. As sketched in figure 5.1, the researchers' "pass-the-food" model considers that each part of the fungus simply receives what it is passed; then it consumes what it needs to survive and passes the rest along. This is very much like a line of people passing along buckets of water on a hot day – or schoolchildren at a traditional long school dining table, such as in the *Harry Potter* movies, passing along the serving plates of food.

In addition to generating realistic macroscopic structures, the researchers find that a wide range of important biological functions also happen to emerge from the model. These emergent properties include the abilities to store food, forage efficiently and even to physically move over large distances. Given the simplicity and generic nature of their pass-the-food rule, the researchers believe that it could play a central role in determining network structures across a wide range of natural systems. Their findings also suggest

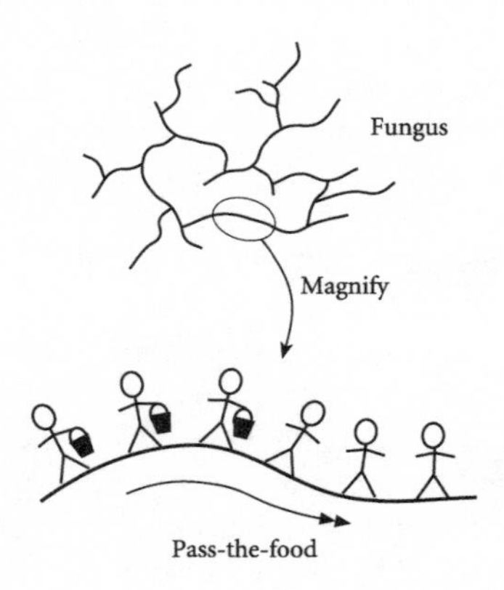

Figure 5.1 Feed the fungus. The fungus transports packets of food, like traffic, along each branch in the network. The researchers' model considers that each part of the branch simply receives what it is passed, consumes what it needs to survive, and then passes the rest along – very much like a line of people passing along buckets of water on a hot day.

that a fruitful method for classifying biological organisms would be by referring to what they do rather than what they look like.

5.5 Looking at money on trees

There is intense commercial and academic interest in understanding the movements in the global Foreign Exchange (FX) market. It is the world's biggest market, and the daily transactions exceed 1,000,000,000,000 US dollars in value, which in turn exceeds the yearly GDP (Gross Domestic Product) of most countries. However it is a formidable task to build such an understanding since the FX market is characterized by a complicated network of fluctuating exchange rates, with subtle interdependencies which may change in time. Indeed it is an excellent example of a real-world Complex System.

In practice, traders talk about particular currencies being "in play" during a particular period of time, as if the FX market were some type of global game as described in chapters 1 and 4. Until very recently, there was no established machinery for detecting such important practical information from market data. However, a joint university-commercial collaboration has recently shown that the construction of so-called Minimum Spanning Trees (MSTs) can indeed capture such important properties of the global FX dynamics. In particular Mark McDonald, Omer Suleman, and Sam Howison teamed up with Stacy Williams of HSBC Bank to uncover the network underlying the movements in the world's currency markets. The research team then showed that this network can be used to detect which currency is "in play" among the world's currency traders.

The novelty of their approach lies in their use of a particular type of network, called a tree, in which there are very few connections. The idea is as follows. Suppose we only have three currencies, such as the Euro, the U.S. dollar (US$) and the U.K. pound sterling (£). An exchange rate is literally an exchange of one currency for another. The shorthand A/B means the amount of currency B which one can buy with one unit of currency A, in which case A is referred to as the base currency. Likewise, it is possible to

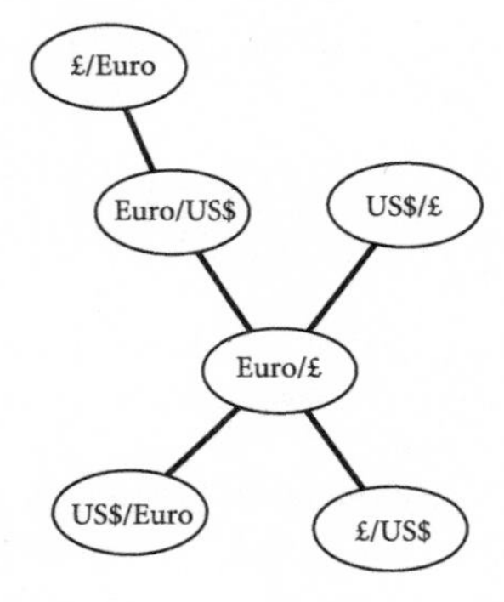

Figure 5.2 Trees growing on money. There are three currencies: UK Sterling (£), US Dollar (US$), and the Euro. Hence there are six possible exchange rates. The appearance of a connection means that the two corresponding exchange rates are moving in a similar way – in technical terms, the two exchange rates are strongly correlated. In this example, the Euro is "in play" in the market.

use one unit of currency B to buy currency A – and this is represented by the shorthand B/A. This means that every pair of currencies has two exchange rates associated with it. Admittedly these rates will almost be the reverse of each other, but they are two rates nonetheless – and they will differ from each other according to whether the prevailing mood in the market favors buying currency A over B or vice versa, or selling B over A or vice versa. The example of three currencies will therefore produce six possible currency pairs – Euro/£, £/Euro, Euro/US$, US$/Euro, US$/£, £/US$. Each of these exchange rates then fluctuates in time in some complicated way. The researchers started off by drawing a network to represent the correlations between the movements of these six exchange-rates. If for example the Euro/£ rate is moving around in a similar way to the Euro/US$ rate, one would say that the two are highly correlated. If the Euro/£ rate is moving around in a way that seems unrelated to the Euro/US$ rate, one would say that the two are uncorrelated. Since there are six different exchange rates, there will therefore be ½ × 6 × 5 = 15 different correlations between pairs of exchange rates. And this is a *lot* of information – too much, in fact, to easily digest. When scaled up to all the world's major currencies, it would become practically impossible for traders to analyze quickly. Hence such a network of correlations in its raw form is of limited practical use.

So the network representing the correlations between all currency pairs has too much information to be very useful in practice. In addition, currency traders would love to have a simple method for deducing whether certain exchange rates are actually dominating the FX market. This would support the popular notion among traders that certain currencies can be "in play" over a given time period. This is where the technical approach of tree construction which the researchers then adopted, proved itself so powerful. The tree – or rather the so-called minimal tree – approach was originally developed by Rosario Mantegna from the University of Palermo and Gene Stanley of Boston University. The trees are minimal in that between n objects, they only contain n-1 connections. Hence as shown in figure 5.2, the three currency situation which leads to $n = 6$ possible exchange rates, has only n-1 = 5 connecting lines between these exchange rates. The tree is constructed as follows: the amount that any two exchange rates are correlated is first represented as a distance, with the most correlated having the shortest distance. Then the most important correlations are picked out so that the network retains its general shape – very much like creating a skeleton structure from a full object. In our three currency example, figure 5.2 shows that the exchange rates involving the Euro as the base currency (i.e. Euro/£ and Euro/US$) are both clustered together near the center of the tree. Using traders' terminology, the Euro is therefore in play. As a check that the resulting tree was actually capturing something meaningful about the market, the researchers created a random FX market by shuffling up some of the data randomly, like a pack of cards. The tree resulting from randomizing the data was very different in character from the true tree, justifying the researchers' claims. Although it is such a simple test – as simple as shuffling a deck of cards – such randomization tests are an extremely valuable tool in the Complexity Science trade, since there is typically very little data available for more conventional tests of statistical significance.

The researchers also looked at how the tree's structure changed over time – as though it were being blown by the invisible wind that shapes the FX markets as a whole. Part of this wind corresponds to the feedback produced by the traders themselves working in the different currencies at different times; part is produced by the economic climate; and part comes from the demand generated by multinational corporations who need to change large amounts of particular currencies. For example, a large motor company like Ford which has lots of sales in the Far East, will often need to turn this cash back into dollars on the FX market, and might therefore create a bit of a breeze themselves. By contrast, you or I going into our local bank and changing money ready for our overseas vacations has no measurable effect on the tree whatsoever. Indeed, we would be very lucky to rustle a single leaf. The researchers found that even though the FX market does change rapidly enough to identify changes in how different currency-pairs are clustering together on the tree, there are links in the tree which can last for several *years*. This remarkable finding shows that there is a certain robust structure to the FX markets. In particular, they sustain themselves like some kind of autonomous machine – a true Complex System.

5.6 Globalization: Fairness vs. efficiency

Let's take another look at a collection of decision-making objects competing for a limited resource – but now we add the twist that some of these objects are connected together. For example, two people might exchange information over the telephone. This research idea was originally examined by Sean Gourley and Sehyo Charley Choe of Oxford University in collaboration with Pak Ming Hui of the Chinese University of Hong Kong. In particular, they carried out computer simulations and developed a mathematical theory to explain their observations, using the crowd-anticrowd picture from chapter 4. Their original motivation for the study was prompted by all the talk in the popular press about globalization. In particular, they started to ask themselves the following questions: is getting connected a good or bad thing? How does increased access to both global (i.e. public) information and local (i.e. private) information, affect the success of both the population as a whole and its individual members? Thinking in terms of future technologies, what are the possible benefits and dangers of introducing communication channels between collections of intelligent devices, microsensors, semi-autonomous robots, nanocomputers, and even biological micro-organisms such as bacteria? These questions will clearly be relevant to a wide range of computational, technological, biological and socio-economic systems over the next hundred years or so.

The researchers' approach was to generalize the binary-decision game mentioned in chapter 4 by adding a certain number of local connections between the objects. As before, the objects in question could be biological (e.g. a population of cellular organisms competing for nutrients), computational (e.g. a grid of software modules competing for processing time), mechanical (e.g. a constellation of sensors or devices competing for communications bandwidth or operating power) or social (e.g. a population of companies competing for business). The researchers' analysis uncovered a rich interplay between the global competition for resources and the local connectivity between objects. For a population with modest resources, they found that adding small amounts of interconnectivity between members of the population *increased* the disparity between successful and unsuccessful people and reduced the mean success rate. By contrast, in a higher resource population they found that low levels of interconnectivity increased the mean success rate and enabled most objects to be successful. At high levels of interconnectivity, the overall population became fairer (i.e. smaller disparity in success rates) but less efficient (i.e. smaller mean success-rate), irrespective of the global resource level.

In other words, they found that the consequences of "wiring up" a competitive population depend quite dramatically on the interplay between the local connectivity and the amount of available global resource. The upshot is that instead of saying glibly that it is good to get connected, we should instead say that "Depending on your priorities regarding fairness and efficiency, it might be good to get connected – as long as you don't get too connected". Admittedly it is not quite as catchy a phrase, but it is certainly more correct.

5.7 The story so far

This brings to a close not only this chapter, but also the first part of this book. We saw that Complexity represents a subtle mix between order and disorder, and that Complex Systems are able to move themselves around between these two extremes without any outside help. Their key ingredient is feedback, which may come in the form of memory from the past or information from other points in space via network connections. We have also seen how we can capture the essence of Complexity using collections of decision-making objects which may, or may not, have network connections between them. With this in mind, we now proceed to look at the applications where Complexity can probably be most useful – and we investigate how we can use the ideas that we have already seen in order to understand the real-world systems of interest. By mixing together elements from our study of collections of objects and of networks, we can then begin to build a coherent, universal picture of a wide range of complex problems.

PART 2

What can Complexity Science do for *me*?

Chapter 6

Forecasting financial markets

6.1 What goes up, must come down: But when?

We'd all like to be able to predict financial market movements. Being able to predict tomorrow's weather or traffic would be an extremely useful skill, but many people would practically sell their souls in order to have an edge in the markets. And with the purchasing power of the average pension diminishing all the time, maybe second-guessing movements in the stock market is something that we'll all be forced to do in the future. There is of course one huge problem: financial markets are complicated, dynamical systems which are continually changing in ways that defy most experts. However, the good news is that they are constantly generating huge streams of data which can be used to cross-check your favorite prediction model – should you be lucky enough to have one.

The main reason for believing that some form of market prediction might be possible lies in the fact that each price movement is actually a real-time record of the aggregated actions of the market's many participants – and each of these participants is effectively trying to win in a vast global market "game". Indeed in its simplest form, this global market game boils down to a binary-decision game of the kind discussed in chapters 1 and 4: should I buy or should I sell? There is therefore reason to believe that prediction models which manage to emulate this underlying multi-trader decision game – in particular, binary decision models of the type discussed in chapters 1 and 4 – could indeed prove profitable.

Prediction in financial markets is fundamentally different from predicting the weather, the outcome from a roulette wheel, or the outcome from tossing a coin. In a market, the individual objects (i.e. traders) are each trying to predict price movements in order to decide whether to buy or sell. The net demand to buy or sell then determines the subsequent price movement. This resulting price movement then gets fed back to the traders, who may use it in their next decision of whether to buy or sell. This cyclic process goes on continually, with price movements being fed back to traders who then make decisions whether to buy or sell. And like all humans, traders can't help but notice what has happened before in the market. They will tend to see patterns – or believe they see patterns – and then react to what they think they see, or what they have heard. In other words, a financial market is riddled with feedback. This feedback leads to new decisions of whether to buy or sell, which leads to a new price, which leads to new feedback, which leads to new decisions, which leads to a new price – and so on.

Such intrinsic feedback does not arise when gambling on a roulette wheel or with the toss of a coin. These objects are made up of molecules – and even though they may appear to behave in a complicated way, they are simply following Newton's Laws of motion. There is no decision-making going on, and hence – unlike the market – the outcome obtained is in no way linked to the predictions of the people who are actually playing and gambling. Likewise, even if everyone had the perfect prediction model of the weather, the weather would still do what the weather does. All that would happen is that everyone would know exactly what to wear the next day. However, this is not true in the markets. If everyone were to be given the perfect prediction model, it would immediately *stop* being the perfect prediction model because of this strong feedback effect. Everyone would use the prediction model in order to decide their next trade, and this would dramatically distort the market. At this point the prediction model would stop working. For example, if the model predicted that stock should be sold, everyone would then try to sell at the same time and the stock would instantaneously become worthless since no one would be prepared to act as a buyer. The upshot is that any prediction model which is too widely known or used will actually hurt the value of the stock held by the traders rather than help them make money.

So, in terms of Complexity Science, financial markets are a wonderful system to look at. They consist entirely of collections of decision-making objects with large intrinsic feedback, and therefore satisfy our main criteria for Complexity. Moreover, the abundance of past and present data means that financial markets provide the most well-documented, and longest running, record of a human-based Complex System in the history of the planet. Hence they are set to play an important role as a test case real-world system in the advancement of Complexity Science – quite apart from their obvious intrinsic interest for commercial purposes.

6.2 The problem with finance theory as it stands – or walks

The method that your pension fund manager or stockbroker uses to manage the risk and contents of your portfolio will always have one huge in-built limitation – no matter how clever he or she is. It can only ever be as good as the model he or she employs to describe the underlying market movements. And since these people are playing with our hard-earned cash, and essentially with our future financial security, we'd better understand more about what model they actually use.

In chapter 3, we discussed a random walk – otherwise known as a drunkard's walk – which is generated by flipping a coin and moving forward or backward one step according to whether the outcome is heads or tails. We saw that the approximate distance moved by the drunkard during a time t, can be written as t^a where $a = 0.5$. This is the *square root* of t. The square root can also be written as $\sqrt{t}$, and so $t^{0.5}$ is just another way of writing this same square root. This means that if we wait 9 seconds, with each second corresponding to one step, then the approximate distance moved by the drunkard away from his starting point will be $9^{0.5}$ steps – in other words, $\sqrt{9}$ steps which is equal to 3. In terms of our earlier filing analogy, the distance moved is equivalent to the change in the shelf position – and in terms of financial markets, it is equivalent to the change in the price.

It turns out that *this random walk represents the standard model of how financial markets move*. And so it is this model which most finance professionals use to try to work magic with our savings. Most importantly, we can see that the main assumption underlying this random walk model of the markets is that the next price movement is best described by the toss of a coin. Heads gives an up movement of the price by a certain amount and tails gives a corresponding down movement.

But that immediately sets off alarm bells in our heads, based on what we have seen in this book. A financial market is a Complex System, and the output of a Complex System will not generally be a random walk. In particular, we discussed in chapter 3 that one of the emergent phenomena in real-world Complex Systems is that they tend to be characterized by a walk that is neither too disordered nor too ordered – more specifically, the patterns that are observed tend to have values of the parameter a which are not equal to the random walk value of 0.5.

We are right to be worried. There is now a significant body of research which confirms that the price-series produced by real-world financial markets are characterized by values of a which differ significantly from 0.5. And here comes the big shock. Not only does the a value differ from 0.5 for any particular market, but the *way* in which it differs from 0.5 tends to be the same *irrespective* of where that market is located. This provides a wonderful example of how emergent phenomena from a Complex System can have universal properties – but more of that in a moment.

At this stage, your emotions are probably mixed. On the one hand, it is fascinating stuff from the point of view of the science of Complexity. But on the other hand, it sounds like terrible news for our pensions in that we cannot trust the standard model of finance.

6.3 A complex walk along Wall Street

So the standard model that most of the finance world uses to calculate how markets move is not accurate. It assumes that the price wanders around according to a coin being flipped – or equivalently, a drunkard walking. However, the actual walk that markets follow is much more subtle. In particular, the market produces values of the parameter a – or equivalently the fractal dimension D discussed in chapter 3 – which are *not* equal to the values for a random walk. (The random walk values are $a = 0.5$ and hence $D = 2$, since $D = 1/a$.) And this discrepancy makes sense: a random walk has no feedback, which is why it can be produced by something as simple as a coin or a drunkard with no memory of the past, whilst financial markets are truly complex since they are riddled with such feedback.

In chapter 3 we saw that fractals have a value of a which differs from the random walk value of 0.5. Fractals appear in the music of Bach and modern jazz, and the coastlines and mountain ranges in our everyday world (see figure 3.3). The reason that such patterns can also appear in the output price-series of markets, is because the markets are continually moving around between disorder and order – just like any archetypal Complex System. And like all Complex Systems, financial markets are able to make occasional forays toward ordered behavior such as a crash, or disordered behavior where no pattern at all occurs, all by themselves. This in turn is entirely in keeping with them being made up of a collection of interacting traders who feed off of global information about past price movements. Financial markets are Complex Systems and they cannot be described accurately by anything other than a theory of Complex Systems. Standard finance theory may therefore appear to work for a while but it will eventually fail, for example in moments where strong movements appear in the market as a result of crowd behavior. And this is far from being a minor flaw since it is precisely these moments when your money is most at risk.

So how do the major financial markets that we all need to worry about, actually behave? Recent research says that the average price-change over t timesteps in most stock markets follows t^a with the value of a being larger than 0.5. In other words, the price in most stock markets follows a walk which is more persistent – more positive feedback if you like – than a random walk. Remarkably, it turns out that many major stock markets throughout the world have similar values of a – typically around 0.7 – despite the fact that these markets are on different continents, have very different sizes and compositions, and have very different net values of trades each day. They may even have very different rules governing when trades can and cannot be made. For example, some smaller markets close at lunchtime, while others do not.

But why should stock markets in such different locations, with different sizes and rules, behave so similarly? In other words, why does there seem to be a universal pattern in the way in which stock markets move? Let's think – the only thing they seem to have in common is that they are dealing with stocks. And that is the key. They are all made up of collections of decision-making objects (i.e. traders) which are continually feeding off past information about price-movements in order to make their next decisions. Whether it is a group of traders in China, New York or London, makes essentially no difference. Instead, it is the way in which these people make decisions based on past information which is important. It is not the actual decisions themselves, but the *way* in which they make them.

Just think back to our model of competing agents in chapter 4. The fact that they all fed off of the same information led to the emergence of crowds and anticrowds – and this same phenomenon is found to emerge irrespective of where that common information is coming from. So it could describe traders in the Shanghai market looking at past Shanghai prices, or traders in the New York market looking at past New York prices. It doesn't matter, as long as the people involved are competing freely, based on the common information which is being fed back to them. The detailed interplay between the crowds and anticrowds will then dictate what is observed in terms of the actual price (i.e. the system's output). In particular, the size of the price fluctuations will depend on the extent to which these opposing groups cancel each other out, and this in turn will vary in time as a result of the feedback in the system.

The fact that we see the same type of price behaviors for stock markets across the world also supports our earlier claim that the overall behaviors of different groups of people can be far more similar than their individual members' traits would have suggested. There is therefore reason to believe that Complexity Science is onto a winner in attempting to describe the collective behavior of human systems.

6.4 Going from the bar to the market

The bar attendance problem discussed in chapters 1 and 4 can be fairly easily turned into a model of financial market movements which reproduces the behavior seen in real financial markets. In particular, the fact that the system can move itself between ordered and disordered behavior as a result of feedback allows it to exhibit a movement in prices which has the same values of a that are observed in real financial markets. This whole topic is discussed in our book *Financial Market Complexity* (Oxford University Press, 2003). Here I will just indicate the issues that need to be addressed in building such a realistic market model.

The main issue concerns the feedback of information. In particular, what is the global information in a financial market? There are clearly many types of information available to traders: for example, the price histories of assets, histories of traded volumes, dividend yields, market capitalization, recent items in the media, gossip, company reports. Any number of these information sources may actually be useful in making an investment decision for a particular asset. However, it is not our interest here to work out which of these information sources is actually useful; instead, we need to know which sources financial market agents tend to make most use of, since we are essentially modeling the population of traders. So if we take a moment to think about what we ourselves see most of, in relation to financial market assets such as stock, the answer has to be its price. The media is full of charts showing recent price movements up and down. Such charts also fill the majority of traders' screens on trading floors. Hence it is reasonable to take the source of global information upon which the traders act, to be based on the past history of prices for the particular market of interest.

Next we have to decide on a method of "encoding" this past history of price movements in a simple way. Inspired by the binary decision games discussed earlier, the simplest alphabet we can use is the binary alphabet of 0's and 1's – just as in the bar problem. A downward movement in price generally occurs when there is a large excess of sellers. Likewise an upward movement in price results from a large excess of buyers. In a similar way to the bar problem, where overcrowding and undercrowding can be represented as a 0 and 1 respectively, we can therefore encode the past history of prices by assigning a 0 to a price movement which is smaller than a given "comfort limit" L (i.e. the market is overcrowded in terms of the number of sellers), and a 1 for a price movement which is larger than L (i.e. the market is undercrowded in terms of the number of sellers). In the context of a financial market, the "comfort limit" L could represent a number of financial or economic variables, which could be either endogenous to the market or exogenous. An endogenous example would be to choose L according to the average market index. An exogenous example would be to choose L according to whether the news of the day was judged to be good or bad for that market. The comfort limit L could also be used to mimic a changing external environment due to some macroeconomic effect: for example, if interest rates are low people may be tempted to put their money into the stock market. Conversely, if interest rates become high then people may prefer to use a risk-free bank account. In other words, L indicates some measure of the attractiveness of stock, or the stock market as a whole, just as it indicates the attractiveness of the bar in the bar attendance problem.

There are, of course, no end of additional details that you could add in. For example: how do financial market agents actually win in the short and long term? What about the fact that each trader or investor will only have a finite amount of money to play with? And how about the fact that some people trade daily, some weekly, and some monthly? It turns out that many of these subtleties, when added into the model, tend to cancel each other out. In particular, the fact that the basic bar model produces a price-series with a similar value of a to real markets, remains robust to these additional bells and whistles.

This brings me to my philosophy for building models of financial markets, and for building models of real-world Complex

Systems in general, which can be summed up in terms of building a paper plane. As we all know, folding a piece of paper into a paper plane can give something that flies, and which therefore captures the essential ingredients of flight, i.e. the uplift cancels the downward pull due to gravity. In short, a paper plane flies for the same reasons that a big commercial jet does. A paper plane is an example of a great model since it is minimal, and yet captures the key ingredient of flight. However some people would not agree – after all, there is no frequent flyer program or meal service. To get such a frequent flyer program we would need to add passengers, and to get a meal service we would need to add flight attendants. But people weigh a lot, they need to sit down, and they have baggage. So we would end up having to add seats, a galley, and hence large jet engines and lots of fuel to help lift it off the ground. In effect, we would end up with such a realistic model that it would actually be nothing less than a full-size commercial jet. So we would have learned how to build an exact working replica of a commercial jet, but we wouldn't have learned anything extra in terms of what it takes for something to fly. For that, we should have simply stuck with our original paper plane and explored different designs. In fact we would probably have learned much more about the intricacies of flight that way. I believe that a similar argument applies to the building of models of financial markets and real-world Complex Systems in general.

Given the paper plane example, it is not so surprising that in order to faithfully reproduce the quantitative features of real market price movements we can use a model as simple as a modified version of the bar attendance problem discussed above. After all, it captures several key ingredients of a Complex System in that it consists of a collection of objects which interact through common information and feedback, and which compete for the best price.

6.5 Look out, we're going to crash

One important emergent phenomenon that the bar attendance model can reproduce is that of financial crashes. These are moments where the market tends to head downward for an extended period. In our Complexity language, this is a classic example of a pocket of order appearing spontaneously out of relative disorder, and serves to confirm why financial markets can be thought of as inhabiting the ground between order and disorder like any other Complex System.

It turns out that there is one simple modification to the bar attendance problem that needs to be made in order to produce realistic crashes. We need to add the feature whereby the decision-making objects (i.e. traders) will only trade if their strategies have been sufficiently successful in making predictions in the recent past. This makes perfect sense – real traders and investors do not trade all the time. Instead, they are continually watching the markets, mentally checking what their strategies would have predicted, and waiting until they are confident enough about their prediction in order to enter the market and make a trade. This simple addition not only allows the traders who are at that moment in the market to form crowds and anticrowds, just as in chapter 4, but it also leads to traders moving in and out of the market in groups. Consequently there can be a rush of traders into, or out of, the market at any one moment. Then once they are in the market, they will tend to either join a crowd or anticrowd. Whenever traders happen to rush into the market in this way, there will tend to be a large increase in the number of trades – hence we say that the market has become more liquid. Conversely, whenever traders are pulling out of the market – which in our model occurs when they become insufficiently confident about their strategies – there tend to be less trades and hence we say that the market becomes more illiquid. Remarkably, and in contrast to common wisdom about markets, our bar model shows that there are several different species of crash. In other words, all crashes are not the same. Instead they fall into different classes – like a taxonomy of crashes.

Thinking more generally, the crashes which occur in real financial markets are actually telling us something interesting about the internal forces present within that market – in particular, the opposing forces due to the crowds and anticrowds. These two forces are usually quite evenly balanced, yielding what looks like fairly random fluctuations in the price. However, crashes are good

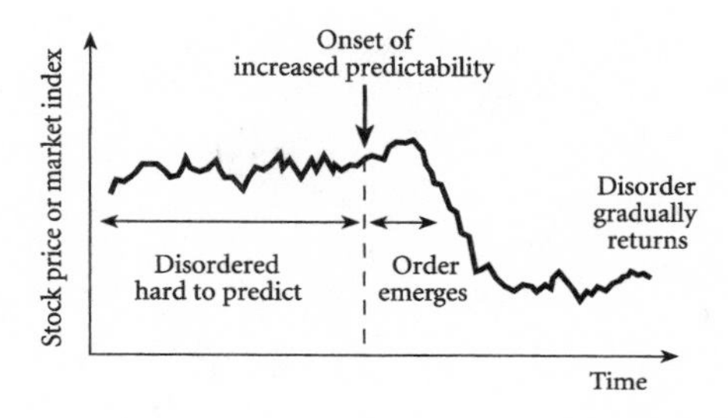

Figure 6.1 Let the bad times roll. The emergence of order from deep within the market means that there is a period of increased predictability prior to the crash.

examples of moments where these forces become unbalanced, hence producing a large ordered movement in one direction. Such large changes are often called extreme events. There are many other examples of such large changes in Complex Systems: punctuated equilibria in evolution, unexpected changes in physiological and immunological levels within the body, sudden congestion in the Internet or vehicular traffic, etc. This ability to self-generate large changes is a defining characteristic of Complex Systems since it allows for evolution with innovation.

Interestingly, the fact that order emerges out of disorder and leads to a sustained crash means that there is the possibility that, with the appropriate spectacles, we could look inside the Complex System and see such a large change coming by detecting this emerging pocket of order. It turns out that this is precisely what is suggested by David Smith's mathematical theory, as discussed at the end of chapter 4. The fact that order starts to emerge prior to a large change means that there is something very concrete which can be detected, and a concrete prediction then made. This effect is sketched in figure 6.1. In the language of chapter 4, what happens as this pocket of order emerges is that the corridors which predict the future movements of the market become narrower and begin to show a definite direction. The upshot is that the market seems to become more predictable before large changes than at other times. And since it is the large changes that investors typically want to know about, because these represent large sources of risk, David Smith's research looks set to become a very important tool – not only for markets, but in all the other application areas described at the end of chapter 4.

6.6 Predicting the future

Let's take a closer look at the whole prediction question, with the focus on financial markets. We have stated above, and in chapter 4, that corridors can be constructed in order to help predict the future evolution of a Complex System. If the corridors are wide and don't seem to indicate a definite direction up or down, this would obviously not be a good moment to make predictions about future price movements. On the other hand, it could be a good time to start trading in market volatility, for example by buying or selling financial derivatives called options. Alternatively, we could just walk into our local betting shop and start spread-betting on the market. By contrast, if the corridors are narrow and seem to indicate a definite direction up or down, this would obviously be a good moment to make predictions about future movements.

But what determines which of these two cases will arise? A financial market walks around in a way which is sometimes disordered and sometimes ordered. On average, this produces a value of our parameter *a* which is between the disordered value of 0.5 and the ordered value of 1. However, at any one moment, the feedback of information to the traders will tend to either act to reinforce the current price trend or to go against it. In the former case, this gives reinforcement of the persistence. The corridors will tend to narrow down and exhibit a definite direction. In the latter case, by contrast, the feedback acts against the current price trend. Hence the market will enter a moment of uncertainty and disorder, in which case the corridors will tend to be wide and have no definite direction.

Building on David Smith's work, Nachi Gupta has recently used real market data to confirm that this type of corridor construction and hence prediction is indeed possible. This finding shows that financial markets are neither continually predictable

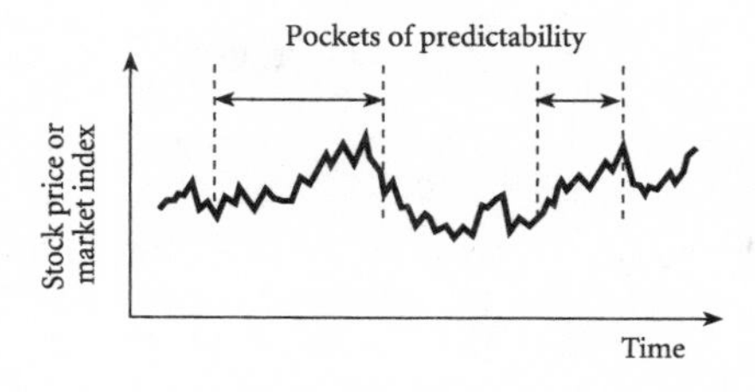

Figure 6.2 Pockets of predictability. The origin of these is the pockets of order that a Complex System will create as it takes itself between the regimes of order and disorder and back again. It is these pockets of predictability that Nachi Gupta and David Smith's mathematical techniques can help identify.

nor unpredictable, but instead show periods where they are predictable (i.e. non-random) and periods where they are not (i.e. random). The basic idea is sketched in figure 6.2. In short, markets exhibit pockets of predictability associated with pockets of order – just like any Complex System should.

In order to produce the corridors and hence identify the pockets of predictability, the research team selects an artificial market model whose price movements closely resemble recent movements in the real market – in technical jargon, they effectively train the artificial market model on real financial data in order to build up a rough picture of the current composition of the trader population within the real market. They then run this artificial market model forward into the future. The range of possible outcomes that are produced then provide the corridors.

6.7 News, rumors and terrorism

Complex systems such as financial markets generate their own forays to order and disorder and back again – like a self-perpetuating machine. They are also typically open to the environment. Financial markets are a perfect example of this, being hit or "kicked" continually by outside news and events, such as companies going bust, earnings reports, unemployment figures, and the outbreak and progress of wars in oil-producing regions of the world. Recent research suggests that most everyday news items have relatively little effect on market movements – but what about major news?

Teaming up with Stacy Williams of HSBC Bank, Omer Suleman, Mark McDonald and Dan Fenn have used the information from Reuters news agency in order to quantify and classify the impact of different types of news on financial market movements. Their ultimate goal is to be able to understand what *types* of news move markets and in what types of ways. In particular, they are attempting to classify news in terms of its different types of impact. Based on an understanding of possible ripple effects across markets, they are working on building general shock-response and vulnerability indicators for different geographic regions or industry sectors, in response to the different types of events which a news agency could report. Their research has already come up with some fascinating results, focusing on the following particular classes of news:

News that is (1) *unexpected* in terms of the fact that it occurs at all, (2) *surprising* in the sense of when it occurs, (3) is *not* related directly to markets, (4) has *never* happened before, and (5) is *true*. As an example of this, the research team considered the shocking 9/11 terrorist attacks in the U.S. In particular, they analyzed the minute-by-minute response of the world's currency markets to the sequence of events as the news unfolded throughout the day.

News that is (1) somewhat *expected* in terms of the fact that it occurs at all, (2) *surprising* in the sense of precisely when it occurs, (3) is *not* related directly to markets, (4) has *never* happened before, and (5) is *true*. As an example of this class of news, the research team considered the 7/7 terrorists attacks in London, and similarly monitored the response of the world's currency markets throughout that day. The fact that the U.K. was a close ally of the U.S. and had hence received terrorist threats since the Iraq invasion meant that such attacks might have been expected at some stage.

News that is (1) somewhat *expected* in terms of the fact that it occurs at all, (2) *surprising* in the sense of precisely when it occurs, (3) *is* related directly to markets, (4) has *never* happened before, and (5) is *untrue*. As an example of this class of news, the research team considered the *rumor* which suddenly started circulating around the world's markets on the morning of Wednesday 11 May 2005, that the Chinese currency would be revalued. The rumor was then denied officially by the Chinese government later the same day. Although such a revaluation had been expected for a while, the timing of any official announcement was completely unknown to the markets in advance – hence such a rumor would have initially seemed highly credible to traders.

News that is (1) *not unexpected* in terms of the fact that it occurs at all, (2) *surprising* in the sense of when it occurs, (3) *is* related directly to markets, (4) *has* happened before, and (5) *is* true. As an example of this class of news, the research team considered the *actual* revaluation of the Chinese currency which occurred later that same year in 2005.

For these particular classes of news, the researchers managed to quantify the response of the currency markets using the network study mentioned in chapter 5. In just the same way as a car mechanic will push down the edge of a car to test the response of the shock absorbers, the team looked at these particular news events to see how the market responded. Their results suggest that markets tend to respond in a remarkably similar way to particular classes of news event, or "shock", from the outside. In the case of terrorist attacks, for example, they found that the global response to the London attacks represented a much milder but similar response to the earlier U.S. attacks. For the case of the rumor and actual revaluation of the Chinese currency, they also found a common pattern of response.

Again using the car analogy, this is consistent with the fact that a particular car will respond in the same way if it is pushed down in the same way. For a car this isn't surprising, since nothing much else is happening to it. But for a market, this is extraordinary. Indeed it represents a huge endorsement of the whole idea that a financial market can be seen as a real entity – a Complex System in its own right, sitting there in the virtual information world. The push isn't real, it is information – and yet the response is real and measurable.

Chapter 7

Tackling traffic networks and climbing the corporate ladder

7.1 Going back to our routes

Traffic can be a real pain, especially when you are stuck in the middle of it. But it creates problems for us even before we have left the house. In particular, anyone who drives to work or does the school-run, is faced with the daily dilemma of "Which route should I take?" Many of us have to deal daily with this dilemma – and many of us try to gain a competitive advantage by making use of our own past experiences and publicly available traffic information. For this reason, these "which route?" problems represent a wonderful example of human Complexity in action: a collection of decision-making objects repeatedly competing for limited resources, armed with some kind of information about the past and present – in particular, drivers repeatedly competing to find the least crowded route from A to B, such that they have the shortest possible trip duration.

Let's suppose, for one idyllic moment, that there are no other cars on the road. Then all we would need to do to get from A to B as quickly as possible is to work out which of the available routes represents the shortest distance. Since we would presumably travel at the same speed on every available route, the route which represents the shortest distance will also be the route with the shortest trip duration. Simple.

The difficulty comes when we add in other cars, and hence other drivers. The more cars there are on a given road, the slower the traffic will move in general. Even if everyone travels at the speed limit, there are just too many things that could go wrong. People tend to slow down if they sneeze, or change radio stations, or look at something by the side of the road – and this gives rise to a chain of events that can end up with that awful stop-start traffic that we all know and hate. Worse still, there might be an accident or some other hold-up that just brings everything to a grinding halt.

The complex patterns which arise in traffic systems result from the interactions between the cars – and these interactions between the cars arise from the decisions and actions of their drivers. Drivers tend to make decisions based on the feedback of information that they are receiving, either through their own personal memories of seemingly similar past experiences or from information about what is going on around them. As a result of this feedback, emergent phenomena such as traffic jams can often appear out of thin air without any obvious cause – just like many financial market crashes also have no apparent cause. This is because traffic systems are constantly shifting between ordered and disordered behavior as time evolves, just like all Complex Systems.

We know that traffic jams are painful. But suppose you have already committed yourself to being on a particular road – there isn't much that you can do, in terms of decision-making, to avoid getting stuck in that jam. Instead, the really important decision-making process actually happened before you took that road: in particular, it was that initial "which route?" question. So let's start by looking at the common "which route?" dilemmas that we frequently agonize over:

Traffic dilemma No. 1: Choosing whether or not to take a particular road

All roads can be thought of as having a certain "comfort limit" L, in the same way that a potentially overcrowded bar or financial market will have an intrinsic comfort limit as discussed in chapters 4 and 6. If the number of cars is larger than this comfort limit, the road becomes uncomfortable to be on. There are typically many other people trying to make the same decision about whether to take the same road or not, and we won't know what the correct decision actually is until it is too late. In other words, we all have to make our decisions and hence take the road or not, and then assess in hindsight whether it was the correct decision based on how many other people decided to do the same thing. So clearly this is the same dilemma as the potentially overcrowded bar or financial market. Therefore everything that we have said so far about these problems – for example, about the emergence of *crowds* and *anticrowds* – will carry over to this road-attendance problem. In chapters 4 and 6, choosing between options 1 and 0 represented choosing to attend a particular bar or not, or choosing whether to buy a particular stock or not. Here it represents choosing to take a particular road or not.

Traffic dilemma No. 2: Choosing between two routes

This dilemma arises, for example, when there are two routes – say route 1 and route 0 – between work and home. Every night we have to decide whether to take route 1 or route 0. Let's assume these two routes 1 and 0 are nominally identical. In other words, it would take the same time to get home using either route, in the absence of all other cars. Then clearly we each want to choose the route which is less crowded – in other words, less cars. So if there are say $N = 101$ of us trying to get home and hence playing the same game, then we would feel we had won if we happened to choose the route with 50 or less cars on it. That would imply that 51 cars had taken the other route, and hence we would have managed to choose the less crowded route. In other words, the worst case that we could possibly experience and yet still be winners, would be to have 50 cars on our road including us, and 51 on the other road. Of course there are much better scenarios for us than this – for example, having only 10 cars on our road and 91 on the other is clearly good. But as long as there is a total of 50 or less cars on our road, including us, then there will necessarily be 51 or more on the other one. Hence we will win. Traffic dilemmas 1 and 2 can be made equivalent, by setting the comfort limit L of each route to be just below a half the total number of competing cars. The "just below" bit is important since we want the two routes to be the same, yet we want to eliminate the possibility that both are under-crowded. Suppose that in our example with 101 cars, we made the comfort limit 51 on each route – then in the case that 50 choose one route and 51 the other, neither route would be over-crowded. Hence everyone wins and we won't get very complex overall behavior. Therefore we should choose a comfort limit of 50.

This second traffic dilemma again represents choosing between two options. In the first traffic dilemma, the choice is to take a particular road or not, and in the language of the bar in chapter 4 it is to attend the bar (option 1) or not (i.e. go home, which is option 0). Within the context of financial markets, there are many similar situations which are equivalent to a choice between two options: to enter a particular market (option 1) or not (option 0), to buy a particular stock (option 1) or not (option 0) – or supposing that you have already decided to be active in a given stock, to buy it (option 1) or sell it (option 0). Going further, you can see that the entire daily activity of any human being can be strung together into a chain of such binary decisions, giving a "tree" of possible outcomes.

Imagine for example, that you are a trader in a financial market and that you have just driven into work. In the period of time since getting out of bed, you have already taken a bunch of decisions in connection with traffic dilemmas – and yet you are now faced with many more:

Should you enter a particular market A or not? Let's suppose you decide "yes" (option 1).

Now, given that you have decided to enter market A, and hence have headed down a particular branch of your daily decision tree, should you be active in a particular stock B within that market or not? Let's suppose you decide "be active" (option 1).

Now, given that you have decided to enter market A, and have decided to be active in a particular stock B, should you buy or sell that stock? Let's suppose you decide "buy" (option 1).

Now, given that you have decided to enter market A, and have decided to be active in a particular stock B, and have decided to buy that stock, should you buy large or small amounts? Let's suppose you decide "large amounts" (option 1).

Now, given that you have decided to enter market A, and have decided to be active in a particular stock B, and have decided to buy that stock, and have decided to buy a large amount, should you then reverse this trade before going home? Let's suppose you decide "yes" (option 1).

And so it carries on throughout the day until eventually you manage to escape from work and head home – only to face a similar set of traffic decisions on the way home. As you can see, these successive decisions may get compounded as the human being in question struggles along his or her particular chain of daily dilemmas. But while the number of different dilemmas that a given person will face in a day will depend on their individual circumstances, the crucial point is that each of these dilemmas is just a copy of the same old binary decision problem – choosing option 1 or option 0 – and hence the same types of emergent phenomena can be expected at each stage.

Traffic dilemma No. 3: Choosing whether to go through the center or go around the outside

Many towns and cities have a road layout which resembles a hub-and-spoke shape. People are therefore often faced with the dilemma of whether to choose a route through the center of the city and risk congestion problems, or to go around the outside and hence risk a longer journey. This dilemma appears to be the same as the two above, since the two possibilities are to go to a potentially crowded place (which in this case means to choose a route through the city center) or not (which in this case means to go around the outside instead). The correct decision will only become apparent after everyone else has made their mind up. However, it is actually a bit more subtle than the previous two traffic dilemmas. In particular, the effective comfort limit L of the city-center option will typically depend on how many connections there are between the outside and the center, and how much congestion therefore develops in the center as a result of traffic taking all of these possible spoke roads. It will also depend on whether there is any monetary charge – such as the congestion charge in London – which will add to the overall cost in terms of time and money.

Thinking back to our discussion in chapter 5 about networks, we can also see an analogy between the hub-and-spoke road network and the hub networks which we said arise in many other real-world settings – for example, social and communications networks. For these reasons, we will devote the rest of this chapter to trying to understand the Complexity that such hub-and-spoke shapes introduce. We will do this in two ways. First we will assume that the flow of objects on the network is behaving relatively simply – in particular, we assume that there is no detailed decision-making being made and hence the whole problem comes down to assessing the effects of congestion in the center. Then in the last part of the chapter, we will combine the decision-making of the bar problem from chapter 4 with this hub-and-spoke shape, to see what you might expect to find lurking around a city center near you. We will also open up the problem to a completely different application area – the social and career ladders that we face, and the dilemma of whether to move around within our existing level or to move up.

7.2 Time is money

We hinted above that going through the center of such a hub-and-spoke network might have several forms of cost. First, there is a cost in terms of the time taken to get through the center. In particular, the congestion in the center itself is determined by the number of cars which pass through the center at any one time. This in turn is determined by a combination of the number of roads (i.e. spokes) which connect the outside to the center, and the number of drivers who decide to take each of these roads. Indeed, one can imagine that drivers who are driving round a ring-road are faced with a bar-attendance-like traffic dilemma as they reach each successive intersection with a spoke-road. For this reason, we will first focus just on the effect of the number of spoke roads. Then we will return later in the chapter to consider the additional complications caused by drivers' decisions. The second cost is a monetary one.

Going through the center may – if we are unlucky – add extra time to our journey *and* cost us real money. Going around the outside may not cost money but, since it is longer in terms of distance, it may end up taking longer if the center is not very congested. The correct decision as to whether to go through the center or around, will therefore depend on how we each balance the importance of time and money. Time is money – but the effective exchange-rate between the two will not only vary between different people, but it may also vary over time for a given person. Interestingly, biological networks such as a fungus (see chapter 5) are also faced with such a dilemma when deciding how to transport food from one side of the organism to another. Man-made road designs are obviously planned in advance – but organisms such as fungi (which are incidentally the largest organisms on Earth) have somehow managed to evolve a balance between centralized supply routes and decentralized supply routes all by themselves. On the forest floor where the fungus sits, there are nutrients such as carbon which need to be transported from one side of the organism to another. If these packets of food – like cars on a road – were all passed through one central point in the fungus at the same time, the congestion could be considerable. Yet the organism survives, and even thrives, without a traffic light in sight.

How does a fungus do it? Nobody knows so far. But this has opened up a fascinating research area which is currently being pursued by Mark Fricker, Tim Jarrett and Doug Ashton to determine how the corresponding transport costs in a biological system dictate the type of network structures that are observed. Not only is Mark Fricker's group well on its way to solving this mystery, they are also gaining valuable insight into possible smart designs for man-made supply networks. Who knows: maybe road planners of the future will simply turn to their friendly neighborhood fungus for new design ideas?

There are many other examples of real-world Complex Systems – both man-made and naturally occurring – where packets of "stuff" (e.g. goods, information, documents, money, datapackets) need to be transported throughout a network structure, and hence the dilemma of choosing centralized versus decentralized routing arises. And just as with the ring road problem, the shortest route between two points may not actually be the quickest and hence best. In particular, if there happens to be a long waiting-time at some particularly busy central hub, then it might have been quicker to send the packet all the way "around the houses" in order to get from A to B. Specific examples include:

Transport systems. In addition to cars on roads and datapackets on communications networks, air travellers and airline schedulers have to decide whether to arrange for a stopover at a large international airport or a regional one.

Supply-chains for goods. A supermarket has to decide whether to have one potentially congested but large central warehouse, or many smaller regional ones.

Information in human organizations. Members of crime and terrorist networks must presumably decide how much information to pass through some central person, which could become a slow or risky option if the person gets overloaded with information such as in a Godfather scenario – or whether instead to have all information delocalized through crime cells.

Administration and policy-making. Multinational companies and government institutions have to decide whether to pass all paperwork through some central head-office for rubber-stamping, or have these pieces of paper wander around between regional offices.

Human biology and health. The growth of a cancer tumor depends on a process called angiogenesis which leads to the creation of nutrient (i.e. oxygen) supply pathways (i.e. blood vessels, like roads) to enable the tumor to continue growing. It is an open question as to the extent to which the tumor balances centralized versus decentralized blood vessel networks. This information could prove crucial in understanding how to restrict the nutrient supply and hence reduce the growth rate of the tumor. Another health-related example of such centralized-decentralized competition, concerns AVMs (Arterio-Venous Malformations) which grow in the brain of many people. These are clumps of abnormal blood vessels which, if the patient is unlucky, may start acting like a new city center through which blood may prefer to flow, instead of taking its normal route through the brain's elaborate collection of very small capillaries. Hence AVMs effectively divert all the traffic through them instead of letting it pass through the delocalized network of capillaries. Consequently, much of the brain becomes starved of nutrients. Recent work by the research group of Paul Summers and Yiannis Ventikos at the John Radcliffe Hospital, has led to an improved understanding of when such AVMs might start switching between decentralized (i.e. healthy) and centralized (i.e. unhealthy) blood flow.

7.3 Creative congestion charging

Anyone faced with designing a network, such as a road layout around a city, is faced with the following problem. Given that each mile of road costs a lot of money, to what extent do you trade off building more shortest paths through the central portion of the network – and hence potentially congesting the center – as opposed to enhancing decentralized pathways around the center, whose downside is that they may make journey times unnecessarily long, or become under-used? Tim Jarrett and Doug Ashton have looked in detail at this problem using a particularly neat mathematical theory to calculate the shortest journey times between any two points on the outer ring of a hub-and-spoke network. Their model correctly accounts for the competition between the following two issues concerning the number of spoke roads that should be built.

On the one hand, one might argue that adding more spoke roads will provide more shortest routes through the center, thereby shortening the average journey time. Hence the answer as to the optimum number of spoke roads to build is simply the maximum number that you can afford. This suggests that one should build as many spoke roads to the center as possible. But on the other hand, the more cars one allows to pass through the center, the more congested it will be. Hence the longer it will take to get from one side of the network to another. This suggests that one should build as few spoke roads to the center as possible.

Tim and Doug have proved mathematically that the competition between these two issues leads to an optimal number of spoke roads that should be built, regardless of the fact that you might be able to afford more. In other words, by building not too many and not too few, the average journey time takes on its smallest value. This is summarized by the sketch in figure 7.1.

In addition to helping understand the interplay between centralized and decentralized flow on traffic networks, Tim and Doug's research represents a fascinating yet novel approach to understanding how biological systems function by focusing on what actually happens on a network, as opposed to just analyzing the shape of the network structure itself. In other words, their work suggests that we should take a new look at biological systems in terms of costs and benefits of connections, rather than in terms of the physical structures themselves. Going back to the more mundane but no less important case of road traffic, their work gives us an important insight into how monetary charges should be levied on toll roads, according to the existing road structure. Their research also tells us something about how road layouts should be designed in order to achieve certain goals for journey times. For example, in the case of an existing hub-and-spoke traffic network, suppose there are 100 two-way roads passing through the center of the city and yet it turns out that having 50 is actually optimal. This could be achieved by

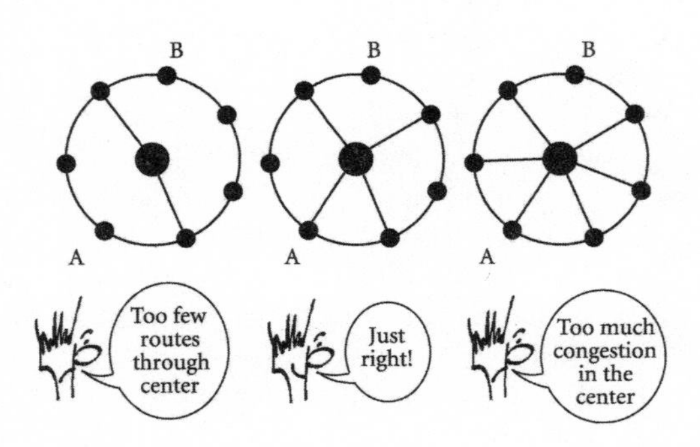

Figure 7.1 Take me to your center? The shortest average journey time is obtained by having neither too few, nor too many routes through the center.

immediately converting them all to one-way roads such that 50 flow out and 50 flow in.

But the most important practical application of their work is likely to be in determining what congestion charge should actually be levied – and more importantly, how. Let me explain what I mean by this. London currently charges a flat fee, regardless of how congested the center actually is when you drive through it. But why should it be a flat fee? Why shouldn't the cost depend on how many other people are travelling through the city center at the same time as you? Indeed, this is an important conclusion of Tim and Doug's work: they find that by varying the congestion cost according to how many other people are also driving through the center, the traffic can be optimized to coincide more closely with the actual number of access roads through the center – or indeed, the number that are actually operational at any given time. So this becomes much more like a market philosophy – instead of more buyers pushing the price up at that moment, more drivers passing through the center will push the congestion charge up at that moment.

7.4 Different shape, same function

The hub-and-spoke network is certainly of interest to most commuters – except if you live in a city like Houston which happens to have two ring-roads around the outside. In other words, Houston has an outer ring, a middle ring and a hub. In fact, there are many other everyday examples of such multiple-ring-and-hub structures: from biological systems through to corporate structures, administrative systems and communications systems. Consider, for example, the lines of communication in a typical hierarchical corporate structure as shown in figure 7.2. It contains many levels of administration, with some connections between people in a given level and some connections between levels. Many readers may themselves already be working in such a structure – and so the question arises: if I want to get a message to someone within another layer, what route should I choose so that I have to involve as few people as possible? More generally, if we

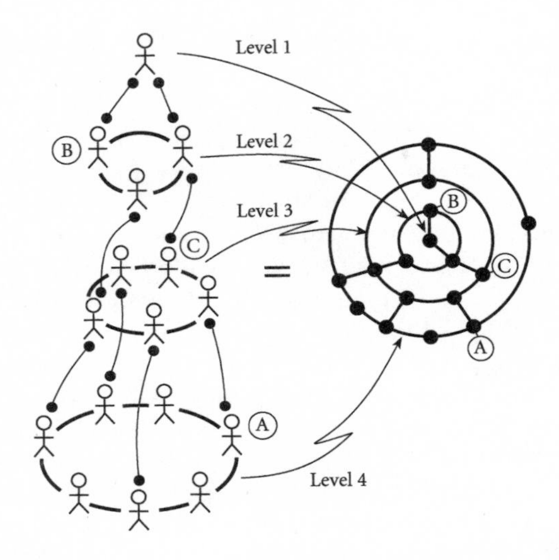

Figure 7.2 Lord of the Rings. A possible corporate, administrative, military or commercial structure of command. If a message needs to go quickly between A and B, and none of the levels is congested, the shortest route is via C. But what happens if level 3 gets too congested? Tim Jarrett and Doug Ashton's mathematical theory helps provide an answer to such questions.

were designing such a corporate ladder, what would be the design that leads to the quickest exchange of information between the most important layers? This could be important not only for corporations, but also for governments and military applications where decisions need to be transmitted between different layers of command as quickly as possible – or in situations where each additional contact represents an additional security risk.

It would therefore be a fantastic breakthrough if one could extend Tim Jarrett and Doug Ashton's mathematical theory to the more complex case of many interconnected rings rather than one. Remarkably, Tim and Doug have recently found a way of doing this for any number of rings. The trick is shown sketched in figure 7.3. By embedding one network in another, the hub-and-spoke of one network then becomes a new "super-hub" – or

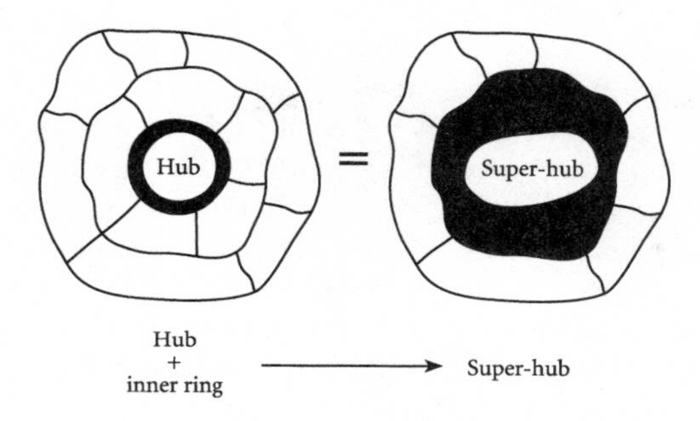

Figure 7.3 It's a maze-ing. By redefining the cost for passing through the central hub, the problem of two rings can be turned into one ring. Repeating this process allows the reduction of a structure with any number of rings, down to a single ring problem which can then be solved using the mathematics developed by Tim Jarrett and Doug Ashton.

so-called renormalized hub – for the larger, multiple-ring network. All that needs to be changed in the mathematical theory is to redefine the cost for going through the center, due to the fact that this renormalized hub now has a more complicated intrinsic structure than before. This renormalized hub can then be combined with other rings to form another renormalized hub which has an even more complicated intrinsic structure and hence a more general cost. Repeating this procedure over and over again, they can turn any multiple ring-and-hub network into a single ring-and-hub. So Houston – you don't have a problem.

Thinking again about the construction of congestion charging schemes, one could therefore imagine having several different central zones – like rings within rings. So imagine that instead of a single central £8 zone in London, there could be two zones: one of £10 and one of £2. Or even better, the exact amounts could be adjusted in real-time, and announced on road-side electronic boards, in order to control the relative amounts of traffic in the two zones as the day progresses.

But Tim and Doug's research has even deeper implications in terms of our understanding of biological systems. Teaming up with Mark Fricker's group working on fungi, they have shown that their theory offers an explanation as to why so many diverse sets of network structures arise in Nature under essentially the same environmental conditions. So let's explore in more detail what they found and why it is so important. We mentioned earlier that there has been a lot of attention paid to the structure of the complex networks which are observed throughout the natural, biological and social sciences. The physics community, in particular, hopes that such networks might show particular universal properties. On the other hand, the biological community knows all too well that a wide diversity of structural forms can arise under very similar environmental conditions. In medicine, cancer tumors found growing in a given organ can have very different vascular networks. In plant biology, branching networks of plant roots or aerial shoots from different species can co-exist in very similar environments, yet look remarkably different in terms of their structure. Fungi provide a particularly good example – different species of fungi can form networks with varying degrees of lateral connections, and yet many exist under exactly the same environmental conditions. But given that such biological systems can adapt their structures over time in order to optimize their functional properties, why on earth do we observe such a wide variety of biological structures under essentially the same environmental conditions?

To answer this, we first need to know a bit about the eating habits of organisms such as a fungus. A primary functional property of an organism such as a fungus is to distribute nutrients efficiently around its network structure in order to survive. Suppose a fungus finds a lump of food somewhere on its perimeter. It needs to transport this food (carbon (C), nitrogen (N) and phosphorous (P)) across the structure to all other parts of the organism in order to feed itself. In the absence of any transport congestion effects, the average shortest path would be through the center; however, the fungus faces the possibility of food congestion in the central region since the tubes carrying the food do not have infinite capacity. Hence the organism must somehow decide how many pathways to build to the center in order to ensure nutrients get passed across the structure in a reasonably short time. In other words, the fungus – either in real-time or as a result of evolutionary forces – chooses a particular connectivity to the central hub. But why should different fungi choose such different solutions under essentially the same environmental conditions? Which one is "right"? Tim, Doug and Mark's work shows that, surprisingly, fungi with very different structures can all be right at the same time. More precisely, they have shown that very different network structures – such as those in figure 7.4 – can share very similar values of the functional characteristics relevant to growth. In other words, the structures in figure 7.4 can both be optimal in terms of transporting food. In particular, it takes the same amount of time in each structure to transport food from one side of the organism to the other.

An important further implication of their work is that in addition to searching for universality in terms of network structure, scientists should consider seeking universality in terms of network function – a message which is echoed by the results of David Smith's fungus model in chapter 5. And in applications like cancer, the ability to say that two apparently quite different-looking tumors may – based on an analysis of the underlying vascular blood network – be identical and hence of equal malignancy, could prove to be an important breakthrough in terms of diagnosis.

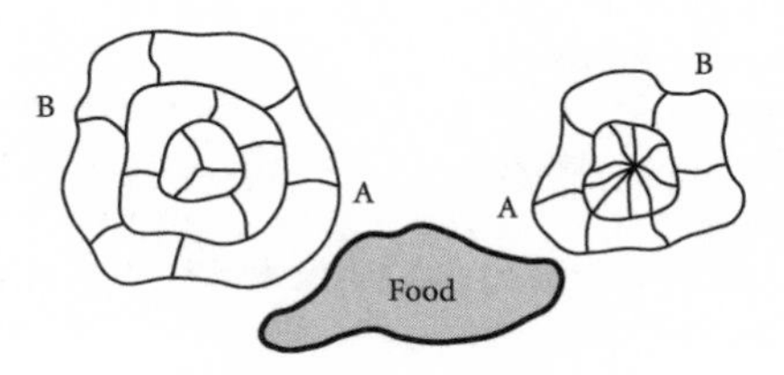

Figure 7.4 We may look different, but we act the same. As a result of congestion at busy hubs, the two structurally different networks shown above have the same functional properties. In particular, food takes the same time to cross from A to B in each structure. Moreover, both structures are optimal in that this time is a minimum for each structure.

7.5 Should I stay at my own level?

So what happens when we put together the hub-and-spoke road network with the decision-making models discussed earlier in this book? In the setting of a corporate structure such as figure 7.2, we might ask ourselves whether we should try to access higher levels of a large organization in order to get a message through from A to B. More generally, should we climb the corporate ladder or is it better to stick to the level we are already at? The big problem is overcrowding. If everybody tries to access higher levels of an organization, or equivalently access the hub or center of a transport network, it will become increasingly congested. In terms of a message, it would be better to transmit it within our own level as far as possible. However the rate of transmission through our own level may be very slow if it includes lots of people. So this is another of our "choose option 1 or option 0" type problems – in other words, traffic dilemma No. 3 in the list given at the beginning of this chapter.

Sean Gourley has looked at the problem of decisions on such hub-and-spoke networks in detail. He allowed the cost of using the central hub to be a variable cost that is dependent both on the number of agents using the hub and the capacity of the hub. In particular, the central hub has a comfort limit given by L, as in chapter 4 for the bar. If this limit is exceeded, then the hub becomes congested and a congestion charge is imposed on the subsequent traffic through the hub. Each agent must repeatedly decide whether to go through the central hub or around the outside – exactly as in chapter 4 when deciding to attend the potentially crowded bar or not. Sean found that the resulting traffic patterns are very rich in their behavior. This richness arises from the interplay between the creation of, and the transition between, particular stable traffic states which arise as the conditions on the network change. His results show that the existence of congestion in the network is a dynamic process which is as much dependent on the agents' decisions as it is on the structure of the network itself. Figure 7.5 summarizes his results, showing that there is an optimal number of connecting roads that should be made available – for most possible values of the congestion charge – such

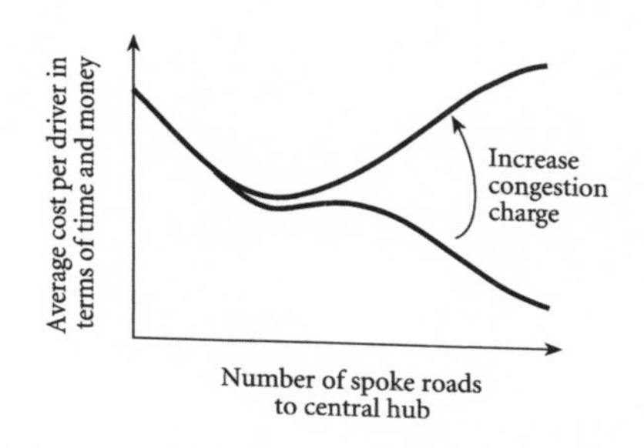

Figure 7.5 Should all roads lead to Rome? Sean Gourley's research results show the effect that additional spoke roads will have on the average cost per driver in terms of time and money.

that each driver spends a minimum amount of time and money getting from one side of the network to the other. This result is therefore consistent with Tim and Doug's findings. In addition to biological systems, Sean's results lend themselves to the many real-world situations discussed earlier – communication across social/business networks, flow of data across the Internet, air traffic, and any other situation in which competing, decision-taking agents have to navigate across a network in which congestion might be a factor.

So let's just take a moment to summarize where we are with traffic, and where traffic research is heading in relation to Complexity Science. Traffic is an interesting example of an interacting multi-particle system (i.e. cars) on a non-trivial topological network (i.e. roads). Many traffic studies have taken the view that cars follow automata-like rules. This is probably a good approximation for dealing with traffic which is already *on* a particular road; however, it does not address the arguably more fundamental question of why those cars, or rather their drivers, chose that route in the first place. For that, we need the multi-agent decision-making models of Complexity Science as discussed above and as highlighted in chapters 1 and 4. Indeed with in-car access to real-time traffic information already available and likely to become more prevalent in the future, an understanding of how motorists' individual decisions feed back onto the emerging traffic pattern and vice versa, is of great practical importance for all of us.

Chapter 8

Looking for Mr./Ms. Right

8.1 The perfect pair

"I've found the perfect person" is a cry you don't hear very often. Why? Because the task of finding Mr./Ms. Right is complicated. Indeed it is so complicated that many people never seem to manage – or think they have managed only to decide later that they have made a mistake.

As many of us might already know to our cost, there are several reasons why finding Mr./Ms. Right is so complicated. First, in order to find that truly perfect person, that person must actually exist. Even if we assume that such a person could in principle exist at some stage in the evolution of mankind, there is no guarantee that they haven't already lived and died before we were born – or will be born after we are dead and buried. Hence it might happen that even if you could trawl the entire planet, you would still never find your perfect match. Worse still, the partner who we didn't think was perfect at some stage in our distant past, may at some later stage seem perfect – but the opportunity has already gone. Second, you actually have to establish contact with this perfect person. The fact that Mr./Ms. Right might live five blocks from you but you never come across them, is a particularly unfortunate but possible event. And just as there are different blocks to explore, there are different towns, countries – and even continents. Third, even if you find your perfect match, you may not be the perfect match for them. Whole sections of literature, comedy and real-life drama, are built around such situations of romantic frustration. Indeed this frustration is reminiscent of the three-file frustration we saw in chapter 2: person A loves person B who loves person C who loves person D who loves person A. Tragic, but true – and very complex.

So those are three major obstacles getting in the way of our search – and here is a fourth, which is also potentially the most complicated: you are not the only one looking. That again sounds obvious, but the fact that you and everyone else is simultaneously looking for that special someone means that we are once again faced with a scenario where we are competing with everyone else for something. In particular, we are each part of a collection of decision-making objects competing for a limited resource, which in this case is the perfect partner. For this reason, looking for a date is as good an example of Complexity as the bar attendance problem, the commute to work, or choosing stocks in a financial market. They are all Complex Systems – hence our interest in spending a chapter looking at this particular problem of pairing.

We spend an enormous amount of time and effort in our lives forming and maintaining relationships of all sorts – including dating and finding friends. Indeed, building relationships is a fundamental human activity. Commercial and political relationships are also fundamental to our Society. For example, as consumers we are all individually in customer-client relationships with particular gas, electricity and phone companies; our employers are typically involved in business partnerships with other companies; and our countries are involved in ever-changing political, strategic and commercial alliances (e.g. the EU and NATO). And, as they say, even "birds and bees do it". Indeed, Nature is awash with various types of mating and grouping rituals. In short, the fact that we are not alone on the planet makes the dynamics of relationships the driving force behind our Social Life on Earth.

But if everyone and everything does it, why does it sometimes seem that we aren't very good at it? Moreover, as people become more sophisticated – or just downright picky – in their requirements for a partner to date, does this mean that Society as a whole is going to be driven to a state where most people are not in a relationship? We are often told that Society is breaking up because we are getting too choosy and hence too ready to break off existing relationships – but is this really so? Similar questions can also be asked of our commercial allegiances to particular products, brands, companies or loyalty schemes – such as airline frequent flyer programs.

8.2 Virtual dating

The key conundrum for many people is whether to hold out for the perfect partner, Mr./Ms. Right, or just make do with Mr./Ms. Right-Now? Richard Ecob and David Smith set about attacking such relationship questions from the perspective of a Complex System – in other words, they have used mathematics and computer simulations to look at how collections of us behave when we wander around within our own social networks looking for the right partner. In addition to helping understand human dating, their results can be applied to situations involving animal mating, businesses seeking out customers, consumers finding a website for their particular needs on the Internet – and even antibodies tracking down viruses. In technical jargon, this problem is very close to the interests of physicists in the areas of reaction-diffusion of particles on networks, and the complex dynamics of far-from-equilibrium systems.

So what exactly did they do and what can they tell us? Their approach was to combine computer modelling with mathematical analysis – and it turns out that their mathematical analysis exploits a remarkable and unexpected connection with the phenomenon of radioactive reactions in nuclear physics.

Their computer model featured an artificial society where relationships could form between "software singles", thereby simulating the scenario of people searching for partners. A spatial network was introduced to represent the social network within which the agents could move and interact. They started the simulations off with roughly equal numbers of men and women. Each person was allocated a list of "personal preferences" which they could use to assess the compatibility of potential partners whom they came across on their travels around their social network. For example, suppose the hub-and-spoke network of chapter 7 were to be chosen as an appropriate social network – this would be suitable for people who have one major meeting place (e.g. work) and then occasionally visit a range of others (e.g. cinema, gym, supermarket). These software people then wander around from site to site, each carrying their own personal list of preferences. These preferences define the person's likes and dislikes – in particular, the list of preferences such as "likes jazz, doesn't like classical music, likes spicy food, doesn't like museums, likes dancing" can be thought of as a person's phenotype, which is a word borrowed from the language of genetics. However, in contrast to the situation in genetics, these preferences are not assumed to be innate – instead they may change in time or arise purely as a result of the person's personal circumstances (e.g. income, lifestyle).

The number of agents per phenotype, the density of agents per social site (e.g. the gym), and the connectivity of these sites, are key quantities in terms of dictating what happens to the population as a whole. Varying these numbers has the potential to change quite significantly the outcome of the computer model and hence the prediction of what happens in real life.

So let's look at the dating arrangements which their computer model used. Suppose some of the males and females meet at a particular social site at a particular time. They then go ahead and compare their personal preference lists – i.e. phenotypes – to see how well-matched they are. If this matching is sufficiently high, they can start a relationship. This pair then stays together an amount of time which depends on how similar the two lists are. In other words, they stay together a longer time if their tastes and interests are similar, and hence if their phenotypes are similar. The model can be easily generalized – for example, the researchers are able to explore the effect of making the people in the relationship gradually change their lists over time so that there is the potential for couples to naturally "drift apart".

They then tracked who is in a relationship with whom, how long the relationship lasts, and for how long these people are then single before meeting someone else. In other words, they were able to measure *all* the things we might secretly want to know about relationships, but have been afraid to ask.

8.3 Radioactive relationships

How long a given relationship lasts is determined by how closely the two preference lists coincide. The fact that relationships might then break up, and people may consequently have multiple consecutive relationships, means that every person's romantic history can be condensed into a simple label such as "two previous relationships and currently single". In other words, any person – male or female – can be labelled as follows:

0S if they have never been in a relationship and are currently single
0R if they are currently in their first relationship
1S if they have had one previous relationship and are currently single
1R if they have had one previous relationship and are currently in another one
2S if they have had two previous relationships and are currently single

and so on. Hence the label "NS" would mean that the person has had N previous relationships and is currently single, while "NR" means that they have had N previous relationships and are currently in a new one, the (N + 1)'th one. In short, anyone's romantic life can be simply represented by a sequence of these labels, 0S → 0R → 1S → 1R → 2S etc., stopping at whichever label describes their current state. Not only does this apply to the software men and women in the virtual dating simulation, but it also describes each one of us. It is strange to think of us as all that way, but it is true for real life as well. So next time you assess your love life, it may be somewhat sobering to see what your label is, following this scheme as sketched in figure 8.1.

What is interesting is the fact that this is also exactly how physicists label atoms which are going through successive stages of

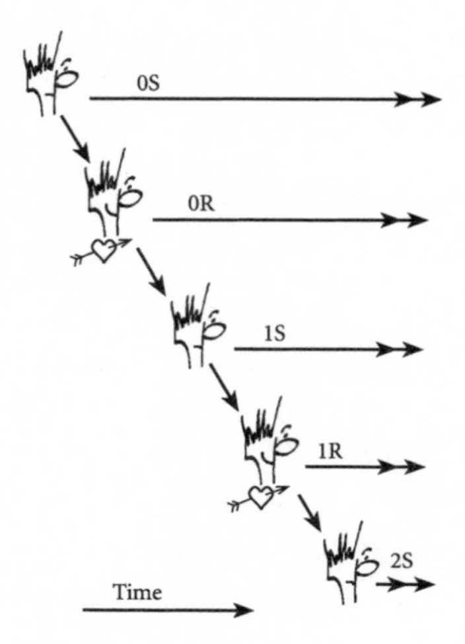

Figure 8.1 The ladder of love. We all start off with zero prior relationships, and single. Hence we start off as "0S". This either lasts forever, or we enter into a relationship and hence change our label from 0S to 0R. If this relationship breaks up, we then have one prior relationship and are single again – in other words, we now have a label 1S. Then if we find a new relationship, we will be 1R and so on.

radioactive decay. In particular, a radioactive atom starts off without having released any smaller pieces through decay. In other words, it is in state 0S. Then the atom starts decaying – in dating terms, it enters a relationship and hence is in state 0R. Then the atom stops decaying, and is hence described as the single state 1S and so on. In this way, Richard Ecob and David Smith were able to describe what happened to the people in their dating model using the language and mathematics of nuclear physics.

Using their simulation and this unlikely mathematics from nuclear physics, Richard and David were then able to build up an intuitive picture of how the ratio of singles to non-singles varies as the conditions change. In particular, they looked at the ratio of singles to non-singles in the long-time limit in order to understand what the eventual mix of the population would be. They could then use this ratio as a measure of how effective multiple-dating actually is. In other words, they could see if the majority of the population eventually became paired off or not. Among the surprising things that they found was the fact that the sophistication of the men and women – in other words, the actual number of preferences in the list of personal preferences – has little effect on the singles-to-non-singles ratio in a large population. Suppose the criterion for forming a relationship is that at least half the preferences should be the same. Even if we double the number of items in the list, in other words the sophistication, a large population will always have a similar fraction of possible matches.

Hence Richard and David were able to show that in this relatively simple set-up it doesn't really matter that we are all becoming more sophisticated by having ever longer lists of preferences. We can still find someone with whom we can form a relationship as long as the lengthening of our preference lists is not accompanied by an increase in our pickiness. In other words, if we tend to form a relationship with people who match at least half of our preference criteria, then lengthening or shortening everybody's list has little effect on whether we will be single on average in the long run. So in this sense, the doom merchants who claim that increased sophistication will lead to a society with an increased number of single people have got it wrong.

By contrast, the average number of agents per site and the way in which these sites are connected together can play a very important role in determining the population's dynamics. Increasing the average number of agents per site and/or the connectivity of the sites has the effect of making multiple dating more efficient. In a real-world context, this corresponds to increasing the rate at which we each meet other people on the network. This can either arise by you moving through the social network yourself or by sitting still and waiting for others to pass onto your social site. The key to the latter in terms of finding a good partner is to sit still at an important place – e.g. near a hub – on the social network.

Let's talk through a particular set of their results, which can be thought of as applying to a set of people who start off dating for the first time as teenagers and then eventually pass through to mature adulthood. We start out with a completely random population of equal numbers of men and women, but with randomly chosen preferences so as to mimic the diversity in a real population. Initially no one is in a relationship, nor have they had any previous relationships. Everyone therefore has the label 0S, and everyone is allowed to move around freely on the social network. Richard and David then allowed the relationships to start forming. The men and women start entering into relationships whose duration depends on how closely matched their preference lists are. Since most of the matches are not perfect, these pairs then start breaking up. Just as in real life, the variation in the length of relationships can be quite large. For example, it turned out that a few couples were indeed so well matched that their relationship lasted a very long time – but these really were the lucky few. The men and women coming out of this first relationship then took on label 1S, until they found someone else who would be suitable to date based on the similarity of the preference lists. And so the whole process went on with everyone heading down the road from 0S to 0R, to 1S etc. at different rates. How far they got down that road in the allotted time depended on whether they had been fortunate enough to come across people with sufficiently similar preferences. But on average, the whole population gradually moved along the same road – slowly but surely as time evolved.

In the long-time limit – in other words, as the teenagers moved though adulthood – the population eventually settled down to have a constant fraction who were in relationships, and some who

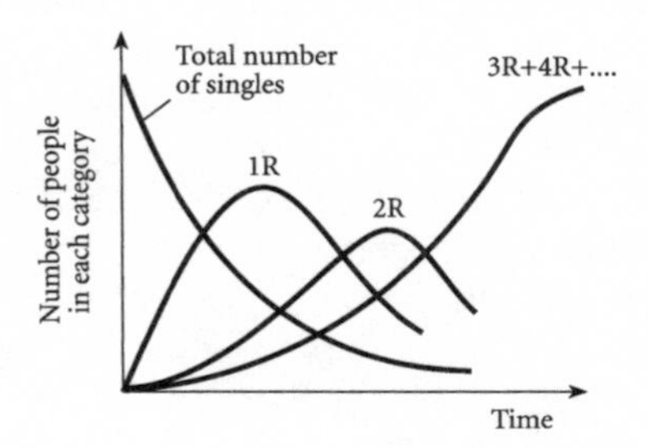

Figure 8.2 Socializing in the City. It turns out that if the number of people per site in the social network is large enough, such as in a lively city, then the number of people in a relationship will eventually tend to become larger than the number of singles.

were single. But just as in real life, it was not always the same people who are single. The population wasn't made up of spinster-types, and married-types. Instead, everybody had their moments of being in a relationship and their moments of being single.

Going further, they found that as long as the average number of people per site is sufficiently high – such as in a lively city with plenty of social life – then the practice of multiple consecutive relationships as in figure 8.1 leads to more people being in a relationship at any one time than the number not in a relationship. These results are summarized in figure 8.2. The implications are that in a lively city repeated dating is a very good thing in that it leads to a population where there are more people in a relationship than single. Although there is intense competition for the right partner, a lively city effectively has a large supply of resources – in other words, a large supply of potential partners in easy-to-access places. Hence the population as a whole tends to do well if the individuals make pairs relatively frequently. By contrast, in a situation such as a rural setting where there is a low number of people per site in the social network, multiple dating actually leads to more people who are single at a given moment in time than in a relationship. This corresponds to the situation of a limited resource – hence the population as a whole does better by *not* making pairs too frequently. Interestingly, these two results for high and low levels of resource mimic what we had discussed in chapter 5 in an entirely different context. There we found that for a competitive population with low resources, making connections was not beneficial. For the situation of high resources, by contrast, making some connections turned out to be beneficial for the population as a whole. So it is essentially the same result – but in a completely different context.

8.4 Here comes your super-date

One interesting modification to this model, is to add "super-daters". In particular, suppose an existing pair can be broken up by the sudden appearance of a more attractive male or female appearing on the scene. In other words, someone appears who is far better in terms of fulfilling the preference list of one of the pair.

We can use these computer models to find out what happens in such a situation by adding people with "magic lists". In other words, they have pretty much everything a woman or man could desire. It turns out that having relationships break up in this way – for example, by meeting someone better at a dinner party – can give rise to a curious effect in the population's dynamics. Such supermen and wonder-women tend to break up many weaker relationships, but can only form one relationship themselves at any one time. Although obviously a destructive process, these break-ups indirectly allow the resulting singles to search around for a potentially better match. Although they may then spend time fruitlessly chasing the super-agent, their travels may allow them to accidentally cross paths with someone who is a better match than their original partner.

This entire discussion, while cast in the language of dating, could equally well apply to businesses looking for clients and vice versa, or institutions looking for partnerships, or people looking for trading partners on the Internet, or search engines looking for word matches on web-pages. For example, you may be a customer of a particular gas company but you suspect deep down that there might be another company out there that would be better for you. It is just that, depending on your threshold, you may have a certain reluctance to want to switch unless someone from that other company actually comes knocking on your door and provides you with the impetus you need. However, there is one big difference between this and standard romantic dating etiquette: in the commercial and political arena, the relationships in question will typically be multiple in the sense that one company is in a relationship with many customers. Polygamy is permitted!

8.5 A more sophisticated dating scene

Teaming up with Tom Cox, Richard Ecob and David Smith have tried out all manner of interesting generalizations of this model. For example, they have tried introducing the idea of births and deaths to simulate a population which is continually replenished from the outside with a supply of single men and women, while taking others permanently out of the dating market. It turns out that this situation gives rise to a remarkably stable state within the population – in other words, replenishment of the population in the form of new blood is good for a healthy dating scene. They have also shown that it can be beneficial for a person who isn't in a relationship to reinvent themselves – in other words, to create a new personal phenotype. Of particular interest is the case where unsuccessful agents try to reinvent themselves by copying successful agents.

David Smith and Ben Burnett have recently taken the mathematical description of the model further by allowing each man and woman to act in such a way as to try to maximize the time that he or she spends in a partnership. The more time spent in a relationship, the more satisfaction drawn. This provides a far more sophisticated set of dating scenarios and situations. They found numerically that there is a highly non-linear relationship between the expected satisfaction level, the threshold for formation of a relationship, and the degree of sophistication of the individual agents. In order to explain this finding, they have developed an analytic theory which depends on the average amount of time which an agent spends in a relationship and the probability of finding a suitable partner on the network. Their analysis can be applied to any network topology, and can be adapted to include biased interactions. For example, it can describe situations where an individual is more likely to meet his or her previous partner. Hmm – I'm sure we all have our own views about such reunions.

8.6 Wolves, dogs and sheep

There are many situations in life where one would actually like to *prevent* pairs or groups from ever forming, and where these models can therefore provide insight. An obvious old-fashioned example is that of human chaperones in which a third-party is introduced in order to keep a relationship from forming (literally). But there is also an interesting example in the medical setting of superbugs and viruses. Suppose a superbug or virus emerges for which no cure is known. Given the right kind of third-party chaperone – in other words, a suitable protein, micro-organism or artificial nanoscale machine – it might be possible to keep such a superbug or virus away from particular defenseless healthy cells, tissues or organs.

This raises an interesting question as to which defense strategy such a chaperone should follow – defensive defense, or attacking defense? This problem is possibly easier to understand using a farming analogy involving wolves, dogs and sheep. Imagine you are in charge of a flock of defenseless sheep, and you know that somewhere nearby there are wolves. You have the possibility of deploying a limited supply of dogs to stop the wolves from reaching your sheep. The question is: what strategy should you train your dogs to follow, in order that they prevent the wolf-sheep pair from ever forming (i.e. to prevent the wolf from killing the sheep)? Should the dogs' focus be to chase after the wolves at the risk of leaving the sheep unprotected? Or should they encircle the sheep hoping that the wolves won't break through? This problem is currently being studied by Roberto Zarama and Juan Camilo Bohorquez in Universidad de Los Andes, Bogota, and we return to it in chapter 10.

Chapter 9

Coping with conflict: Next-generation wars and global terrorism

9.1 War and Complexity

There are several ways in which people can form groups. In particular, group formation can be unintentional – for example, certain subsets of people may just happen to be following the same strategy, as in the case of the crowds and anticrowds of chapter 4. Or it can be intentional – for example, where individuals are looking for a partner, as in the dating scenario of chapter 8. In this chapter we turn to think about what such groups might actually do once they have formed.

Groups of people can be violent. History is riddled with examples of crowds initiating tortures, executions, riots and attacks. But perhaps the most violent act of all is the collective human activity of warfare itself, in which several groups of people simultaneously fight for some kind of gain. This competition to gain something can be seen as a competition for some kind of limited resource – just as drivers effectively fight over space on a potentially crowded road, bar-goers effectively fight over seating in a potentially crowded bar, and traders effectively fight for a good price in a financial market. Even when dating, we are effectively fighting – albeit in our own group of one – for the rare commodity of the perfect partner. Likewise in wars and human conflict, there is a fight going on for a limited resource, where the resource in this instance corresponds to land in a given country or part of the world – or political, social and economic power.

But if wars are just another example of collections of humans competing for some limited resource – as in the traffic, bar, or markets – then they are also examples of Complexity in action. The fascinating thing is, therefore, that we may be able to understand wars in terms of such Complex Systems analysis. This would in turn suggest that the way in which wars evolve has less to do with their original causes and more to do with the way in which humans act in groups. Indeed, it is often reported in wars that many of the people doing the fighting do not actually know why the war started in the first place, nor have they been told the concrete objectives which the war's originators wished to achieve. A good example of this is the on-going guerilla war in Colombia, South America, where there are several different armed groups. It turns out that many of the present combatants are unaware of their own side's overall agenda – they simply want to "beat the others".

In chapter 6 we mentioned how stock markets in very different parts of the world tend to show the same fractal patterns in their output price-series. We attributed this to the fact that any particular market's movements simply reflect the activity of its traders – and, irrespective of their origins or nationalities, traders are just human beings making decisions based on the information being fed back to them. Even though they may occasionally respond to particular exogenous events in their own environment, most trader activity is endogenous in that they are reacting to their own collective past decisions – as in the everyday scenarios of chapter 4. So why shouldn't the same reasoning be applied to explain the dynamical evolution of wars? In this chapter we'll look at some very recent research which offers strong support for this idea of a *universality* in wars, as a result of generic human activity.

Wars used to be simple – or rather, it used to be relatively simple to understand the mechanics of how wars were fought. There were several reasons for this. First, there were typically only two opposing forces – for example, the Saracens and the Crusaders. Just like God and the Devil, or Good and Evil – although, of course, which was which depended on which side you were on.

Second, the weapons which each side had available were similar. In other words, the same technology was available to each side. Third, the sizes of the two armies were usually fairly similar. For these reasons, each side would be willing and ready to fight in a similar way to the other. This led to very conventional warfare. Given such symmetry, the method of confronting the enemy was typically to line up all of your own army on one side and let your opponent do the same on the other side. Then at dawn you would simply try to knock the stuffing out of each other. There was very little element of surprise. When things became less symmetric – for example, when an army found itself fighting on terrain that was much better suited to the other, or with far fewer soldiers – strategies would become more important. However, the basic underlying symmetry typically remained.

As imperial and colonial histories developed warfare became less symmetric. In such asymmetric situations the sizes of the two armies involved are no longer similar, nor is their respective technology and weaponry similar. Indeed, civilians might also be ready to join in the fighting using hand-made weapons – hence the "enemy" for a conventional army might become relatively unorganized, possess more provisional equipment and weaponry, but be far deadlier because of its sheer numbers and potential indistinguishability from the civilian population. Such was the case in Vietnam, Northern Ireland and Afghanistan – and more recently in Iraq and Colombia.

In addition to this increasing asymmetry over time, recent wars have tended to involve more than two sides. Up to the end of the Cold War there was a definite sense of "war games" being played out between just two players. Irrespective of whether they were similar in size and/or technological advantages, side A would react to what they thought side B would do – or, as in chess, side A would make a pro-active move in order to prevent side B from doing something potentially advantageous. As a result the equivalent of a stalemate typically arose. Each side only had to work out what the other side might do, or had just done, in order to know what it should do. So war, while of course still horrific, was relatively simple – like any game involving just two players.

Wars involving three or more players – be they insurgents, guerillas, paramilitaries or national armies – are far more complicated. Just as we mentioned in chapters 2 and 8, frustration can arise. If A hates B, and B hates C, does that mean that A must therefore like C? Not necessarily. Hate is many-sided, just as love can be. Again we only need to think about the ongoing insurgencies in places such as Colombia, where there are many armed groups, to see the potential complications. A sides with B, B sides with C, but A hates C. Therefore A starts to fight B so as not to favour C – and the whole process carries on. Indeed, such frustration may be why many modern conflicts seem to go on and on without reaching any definite conclusion. *Two is company, three is Complexity* – and as we have seen throughout this book the dynamics and time-evolution of such "many-body" situations is very complicated. And like other Complex Systems, any given war will have the unfortunate ability of being able to generate extreme events all by itself – just as a market can spontaneously produce crashes and the traffic can spontaneously produce jams.

The evolution of such many-player wars can be thought of as an ecology in which there are many co-existing species. In Colombia, for example, the war involves several guerilla groups, terrorists, paramilitaries and the army. But what makes such wars so complex is that nobody knows exactly how these different species will interact at any one moment. For example, if a guerilla group from army A meets a guerilla group from army B, will they fight? Or will they choose to collaborate by ganging up on a nearby army C? Or will they simply ignore each other – or maybe even consciously avoid each other? And how does all this change in time?

Adding to this complexity is the fact that a modern war such as that in Colombia, and to a lesser extent Afghanistan, will take place against a backdrop of illicit trade such as drug trafficking. This activity provides "food" for some of the groups in the form of money for buying supplies and weapons, and thereby helps feed the war as a whole. Just as in the case of a growing fungus or cancer tumor, Colombia in particular has multiple nutrient supply-chains corresponding to (1) the flow of cocaine along supply-routes from the jungle to the cities, and then to the U.S. and Europe; (2) the flow of cash which this cocaine supply then generates; (3) the flow of kidnapped victims from the cities back to the jungle. So just like the fungus which thrives in the forest, or the cancer tumor which thrives in the host, these armed groups are fed by a rich source of nutrients which allows them to self-organize into a reasonably robust structure – and just like the fungus or cancer tumor, this makes the problem very hard to get rid of.

Wars are fought by collections of people who organize themselves into groups – for example, units, factions, brigades and armies. Decisions are then made by individuals and groups, and these decisions lead to actions and events which propagate the war itself. Most importantly, these decisions are in turn affected by past and present events – in other words, there is feedback. As a result of this feedback, the objects in question then interact in a potentially complicated way. The net result is a complex history of attacks and clashes during the lifetime of a given war, with each such attack or clash generally producing injuries and fatalities. Although of course shocking, the important point from the perspective of Complex Systems is that such casualty data can therefore be used as a measure of ongoing activity in the war.

Very recently, Mike Spagat of the University of London, together with Jorge Restrepo and his team at CERAC in Bogota, Colombia, carried out a detailed analysis of the attack and casualty data in a number of ongoing wars, including Iraq and Colombia. To their surprise, they uncovered patterns in this casualty data. Even more remarkably, they found that the patterns for the two seemingly unrelated wars in Iraq and Colombia, are currently the same. This suggests that – regardless of the origins of these two separate wars or the ideologies behind them – the insurgent groups in each war are now effectively the same in terms of how they operate.

Mike and Jorge's research team have also managed to develop a mathematical model based on group-formation, which describes a likely scenario for how the insurgent groups are operating. The model gives very good agreement with the observed patterns in Iraq and Colombia. This good agreement implies that the mechanism described in the model has managed to capture how the real insurgent forces are currently operating in these countries. In particular, the model suggests that the insurgencies in both wars involve a loosely connected soup of so-called "attack units" which are continually combining and breaking up over time. When a given attack unit undertakes an attack, it will tend to create a number of casualties proportional to that attack unit's strength. Hence the distribution of strengths of these attack units should reflect the distribution of casualties which arise in the war. And this is exactly what they have found – indeed, the model reproduces almost perfectly the patterns which they have uncovered from the casualty data. Mike and Jorge's findings are so remarkable that we will discuss them in detail in Section 9.3. But first we will go back to the early part of the twentieth century to understand the significance of such patterns.

9.2 The Law of War

Our story starts with Lewis Fry Richardson who worked as an ambulance driver in World War I. Richardson collected together the total casualty figures from each war that took place between 1820 and 1945 – and when he plotted them on a graph he discovered something which is quite amazing. But before I can tell you what he found we need to think about what graphs usually look like.

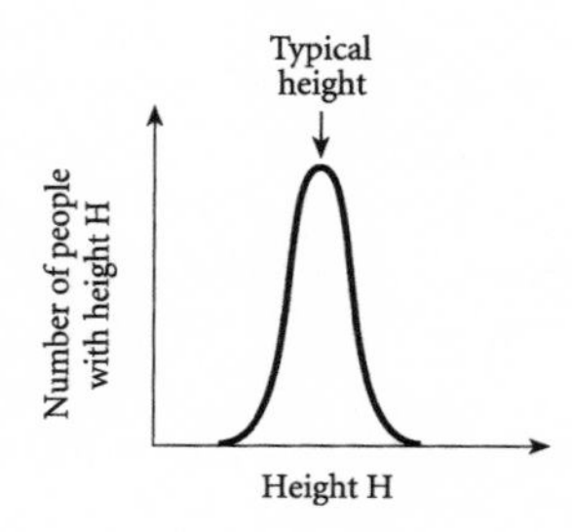

Figure 9.1 What it means to be "Normal". A graph showing the distribution of heights of people. The number of people at a given height H is shown, for all possible heights H. Since no one can possibly be ten feet tall, and no one is less than one foot tall, it makes sense that the shape will look peaked at intermediate values and then fall off to zero on either side. Such a curve is called a bell-curve or "Normal" distribution.

Suppose we know the heights of everyone in our street, or city, or country – or even the world. If we then make a graph of the distribution of these heights we will get something like figure 9.1. Since no one is ten feet tall, and no one is less than one foot tall, it makes sense that the curve will rise up and then drop back down again. This also means that there will be a peak – like the top of a mountain is a peak. This peak occurs at the height which describes the largest number of people – in other words, it represents the typical height. Everyone has a height close to this value. The important point for our story is that there exists such a thing as a typical height. So imagine that we subsequently had to guess someone's height without ever having seen them. If the peak of the curve in figure 9.1 occurs at 5 feet 10 inches, and the spread around this value is eight inches, then we would be pretty safe in suggesting that this unknown person's height is 5 feet 10 inches give or take about eight inches.

Many other curves would also look like this one – for example, the speed of cars on a road. In fact, in both cases the shape is this same bell-shape, or so-called "Normal" distribution. And there is a reason why: the average in each case is dictated by something structural and pre-determined, while the spread in values around the average is due to environmental *ad hoc* reasons. As far as the heights go, a person's body has an implicit reason based on genes and inheritance to grow to a certain height. Then, if the person has an extreme oversupply or undersupply of nutrition, they will probably end up somewhere just above or below this value. The same idea holds for traffic: there is some pre-existing speed limit on a given road which tends to control the average speed of the cars; on top of that, there are everyday environmental and behavioral reasons why individual drivers may drive slightly above or below this value.

Richardson might have expected that, in a similar way, there would be a "typical" size of a war, with a typical number of casualties and a spread around this according to particular circumstances. But this is not what he found at all. Instead he found that the number of wars with a given total number of casualties N, decreased as N got bigger. In other words, the curve did not have a peak as in figure 9.1. The number of wars with very few casualties turned out to be the largest part of the curve, and then the curve just decreased. Maybe this isn't so shocking – but what he found next certainly was. Instead of just decreasing in any old fashion, he found that when he drew the graph in a certain way, the shape was essentially a straight line. In particular, when he took the so-called logarithm of the number of wars with a total of *N* casualties and plotted it against the logarithm of the number of casualties *N*, he got a near-perfect straight line as sketched in figure 9.2. (A "logarithm" is what you see on a calculator as "log" or "ln". Richardson of course didn't have a calculator – instead he had to use those old log-tables that they used to hand out in school).

This finding is quite remarkable. It is hard to think of anything less ordered or regular than war. Wars have different causes, are fought by different people in different parts of the globe, and seem so horrifyingly unique that it would appear impossible that any significant similarities would arise. Yet what Richardson found is not just a statement of similarity in words – it is a precise mathematical relationship. Going further, it is a law – a mathematical law of war. My grandfather told me of the horrors that he lived through in the trenches in France in World War I, and would have

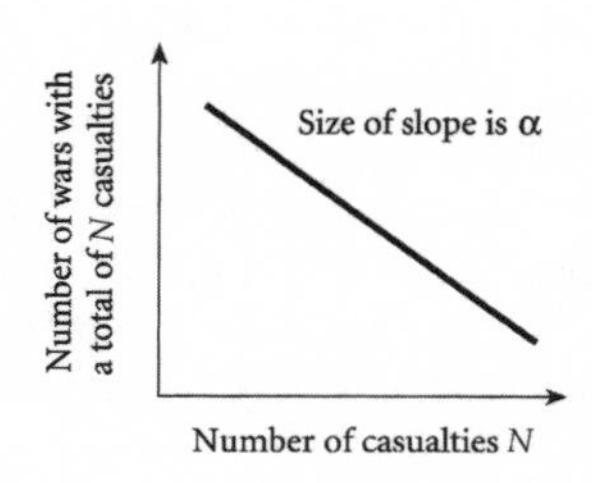

Figure 9.2 Wars are not "Normal". Instead, the distribution is a so-called power-law. The term "power-law" has nothing to do with physical strength. Instead it describes the fact that when plotted on a log-log scale as shown schematically here, the number of wars with a total of *N* casualities against the number of casualties *N* is a straight line. Equivalently, the logarithm of the number of wars with a total of *N* casualties plotted against the logarithm of the number of casualties *N*, is a straight line.

been the first to say that there is no law whatsoever in war. But that would be wrong. World War I is one of the points on that straight line, and so are all the other wars from that era of history.

But why would such a law of war exist? Richardson didn't really know – but fortunately we can now explain his remarkable finding using Complexity Science. The clue lies in the so-called *universal pattern of Life* which we discussed in chapter 3. There we found that Complex Systems live between disorder and order, and tend to produce particular types of fractal patterns that look like t^a. The same holds here in that the number of wars with a total of N casualties follows a mathematical relationship like $N^{-\alpha}$. This mathematical relationship is more generally referred to as a power-law, and α is just a number in the same way that a is just a number. Now imagine we grab a calculator and work out the logarithm of the number of wars with a total of N casualties, and plot it on a graph versus the logarithm of the number of casualties N. It will look like figure 9.2. The reason is simple: the logarithm of the number of wars with a total of N casualties is equal to the logarithm of $N^{-\alpha}$. But the logarithm of N raised to the power $-\alpha$ is just $-\alpha$ multiplied by the logarithm of N, which is $-\alpha \log N$. That is the way logarithms work. So this means that the logarithm of the number of wars with a total of N casualties, plotted against the logarithm of the number of casualties N (i.e. $\log N$) is just a straight line with a slope equal to $-\alpha$. In fact, Mark Newman of the University of Michigan has recently shown that Richardson's straight line – in which each point corresponds to the total number of casualties in a given war – has a value of α of around 1.8.

The fact that wars follow a power-law has some important consequences as compared to the bell-curve distributions typified by figure 9.1. First the good news. The most frequent size for a war will be one with fewest casualties, unlike the case of people's heights. Now the bad news. Very deadly wars and attacks with many casualties will occur – rarely, but they will occur. This is unlike the case of heights, where the chances are zero that someone will be taller than ten feet. For this reason, planning for wars is inherently a complex task. House designers can happily put the height of an entrance at something less than ten feet knowing that such a tall house-buyer will never appear. They can also put step heights above one inch, knowing that such a small person will never appear. However, the presence of a power-law means that this type of assumption will not work for wars. Unlike the bell-curve, the distribution of wars predicts that future conflicts can have an extremely wide range of casualties. This would suggest that instead of planning for the typical future war, planners should indeed plan for the worst case.

The significance of all this for our story of Complexity is as follows. We have said that a Complex System is a collection of objects interacting in some potentially complicated way in the presence of some kind of feedback. But this is also exactly what wars are. We have also made a big point throughout this book of saying that human systems involving a collection of decision-making objects, which are competing for some kind of limited resource as in the bar attendance problem, the traffic, and the market, are all excellent examples of a Complex System. But this is again exactly what we also said about a war. Whether that limited resource is land or power doesn't matter. When it comes down to it wars are "just another" collective human activity. There is no "invisible hand" or central controller to decide who wins. Hence there is no typical war with a typical number of casualties. Instead the system of adversaries just fight it out and the whole thing takes on a life of its own – just like every other Complex System on the planet.

9.3 The universal pattern underlying modern wars and terrorism

Aaron Clauset and Maxwell Young of the University of New Mexico recently took another look at the work of Richardson – but this time in the context of terrorism. In fact they repeated what Richardson did but used instead the number of casualties per terrorist attack rather than the number per war. What they found was equally remarkable to Richardson's original results. Despite the fact that terrorist attacks are typically well spread out in time and in space – in other words, they occur quite rarely and are spread out all over the planet – they found that when they plotted the logarithm of the number of attacks with a total of N casualties versus the logarithm of the number of casualties N, exactly as Richardson had done, they *also* saw evidence of a power law. In other words, the number of terrorist attacks with a total of N casualties varies according to $N^{-\alpha}$. When they restricted themselves to terrorist attacks occurring in (what were then) non-G7 countries, they found the value of α to be 2.5. For terrorist attacks occurring in G7 countries, they also found a power-law, but with α now equal to 1.7.

This is where Mike Spagat and Jorge Restrepo enter the picture. Together with their collaborators in Bogota, Colombia – in particular, Oscar Becerra, Nicolas Suarez, Juan Camilo Bohorquez, Roberto Zarama and Elvira Maria Restrepo – they built and analyzed a huge, detailed dataset for the twenty-plus-year war in Colombia. Then they did the same for the war in Iraq, building on the database of the Iraq Body Count team. If they had wanted to, the researchers could have just added together all the casualties in each war, and then added these two datapoints – one for Colombia and one for Iraq – to Richardson's curve in figure 9.2. However, they instead did something far more interesting and unique.

What Mike Spagat, Jorge Restrepo and the rest of the team did was to pursue the following line of thinking: wars follow a power-law, and wars are a human activity. But given that a war is generally made up of lots of smaller battles or clashes, like "wars within wars",

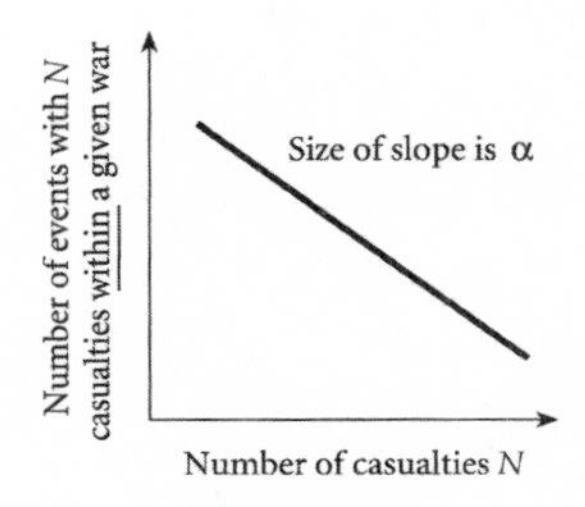

Figure 9.3 Wars within a war. Pattern of casualties from the events within a given war, such as Iraq or Colombia. The graph shows a log-log plot of the number of events with N casualties within a given war, against the number of casualties N.

would we also see a similar pattern emerging within a single war? In other words, can a single war be seen as a set of wars-within-wars?

This is exactly what they found. Despite the very different origins, motivations, locations and durations of the wars in Iraq and Colombia, they found similar power-law patterns in the casualty figures for the events within each war. This result is sketched in figure 9.3. Their finding is remarkable not only because of the different conditions of the wars, and their very different locations, but also their different durations. The Iraq war is basically being fought in deserts and cities and, at the time of writing, has only been going on for a few years. By contrast, the guerilla war in Colombia is mainly fought in mountainous jungle regions, and has been ongoing for more than twenty years against a fairly unique back-drop of drug-trafficking and Mafia activity.

What Mike and Jorge's findings suggest is that the way in which such modern wars unfold as time goes by has less to do with geography or ideology and much more to do with the day-to-day mechanics of human insurgency – in other words, it has to do with the *way* in which groups of human beings fight each other. And this is *exactly* the same kind of common feature that we found in financial markets in chapter 6. Despite their very different locations, operating rules and ages, two markets can have very similar values of their fractal parameter a (which is analogous to the size of the power-law slope α in the present context). The reason? It is all down to human nature. When left to their own devices, without any "invisible hand" or central controller, human groups interact in such a way as to produce markets with similar characteristics, and wars with similar characteristics. This is because a collection of humans who are competing for some limited resource is an excellent example of a Complex System – and Complex Systems show certain levels of universality. And wars, like markets and traffic, are all just examples of Complex Systems.

They then took their analysis further, looking at how the war is evolving in time. In other words, they chopped up the length of the war since its beginning into little sections and found that the data in each piece also followed a power-law. They then deduced the slope of the power-law for each piece. Their result is sketched

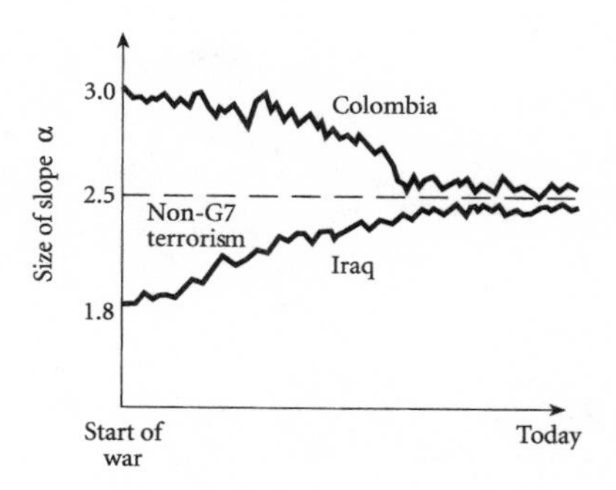

Figure 9.4 The future of wars? The two modern, but very different, wars in Iraq and Colombia seem to have evolved in such a way that they currently have the same form. And this form in turn matches the pattern of terrorist attacks in non-G7 countries.

in figure 9.4. Remarkably, the size of the power-law slope in each war has crept toward 2.5 which is exactly the same as the coefficient for global terrorism in non-G7 countries. This suggests that both these wars and global non-G7 terrorism currently show the same underlying patterns and hence character. This in turn suggests that the insurgent forces underlying these modern wars and terrorism, are now effectively identical in terms of how they are operating. You might think this is good news since resolving one of the conflicts could then give a clue about how to resolve the rest. On a more pessimistic note one could of course say that we won't resolve any of them without resolving all the rest. In some sense, it is all part of one big ongoing war.

9.4 A Complex Systems model of modern warfare

But what is so special about an α value of 2.5? In other words, why does the value 2.5 emerge for things as supposedly unconnected as Iraq, Colombia and non-G7 terrorism? The answer lies in the human activity of forming groups.

The mathematical model developed by Mike Spagat and his team puts forward the idea that any modern insurgent force – whether in

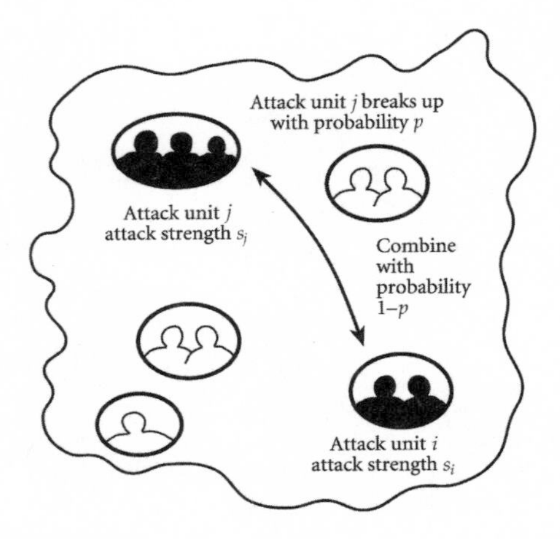

Figure 9.5 Breaking up and getting together: a model which reproduces the power-law pattern observed in the modern ongoing wars in Iraq, Colombia and global terrorism in non-G7 countries.

Iraq, Colombia, or a terrorist organization – operates as a network of fairly self-contained units which evolve in time. They call these units "attack units". Each attack unit has a particular "attack strength" which indicates the average number of casualties which will arise in an event in which this attack unit is involved. As the war unfolds in time, these attack units either join forces with other attack units or break up. In the real war, joining forces or breaking up would probably involve a decision process – hence one would ideally invoke a decision-making model similar to the bar-attendance and "which route?" problems of chapter 4. In other words, we have said that a war is a collection of people competing for limited resources – so a model of whether to choose option 1 (i.e. combine with another attack unit) or option 0 (i.e. break up) might make sense. However, such a combination of models would be difficult to analyze mathematically. Instead, Mike and Jorge found that they could use a much simpler description of insurgent decision-making, and yet still explain the observed data. In particular, the researchers assumed that the attack units effectively use a coin to make decisions about whether to combine or break up. In particular, Mike and Jorge assumed that attack units join together with a given probability 1-p, and break apart with a probability p. They then allowed this process, whereby attack units break up or combine, to carry on indefinitely. To their surprise they found that the insurgent force reached a kind of status quo in which the distribution in the number of attack units with a given attack strength followed a power-law. Since each attack unit will produce an average number of casualties equal to its attack strength in any given event, this distribution also represents the distribution of the number of events with a given number of casualties. So this agrees nicely with the real casualty data – but the surprises don't stop there. Remarkably, the value of the power-law slope which emerges from the model is 2.5, which is the same value of α as that obtained for the real wars and non-G7 terrorism in figure 9.4. This is an incredible finding.

Why should a power-law arise from such a model? We already discussed how the real war is a Complex System – but what about the model? We saw in chapter 3 that feedback is needed to produce a power-law – so where is that feedback coming from in Mike and Jorge's model? It turns out that it arises as follows. The precise distribution of attack units that are available for breaking up or combining at a given moment in the war will depend on what has been happening to the soup of attack units leading up to this moment – and this, by definition, means that there is feedback from the past. As a result of this feedback the distribution of attack units, and hence casualties, will neither be completely disordered nor completely ordered. Instead, it gives us something which is complex in exactly the same way that the slightly sober drunkard's walk in chapter 3 was complex. That example gave us a complex walk which was neither completely disordered, like a true drunkard, nor completely ordered like a sober walker – it is a fractal. Here we have the equivalent – but in terms of group-formation, in a space containing insurgents.

The upshot of all this is that Mike and Jorge's findings offer a novel "Complex Systems" interpretation of the evolution of modern wars and terrorism. In particular, their results in figure 9.4 suggest that the Iraq war began as a conventional confrontation between large armies, but continuous pressure applied to the

Iraqis by coalition forces broke up the insurgency into a collection of attack units. In Colombia, on the other hand, the same end result has been arrived at in the opposite way. In particular, the guerrillas in the early 1990s were unable to join up into high-impact units – hence the attack units were all very small. But since then they have gradually been acquiring comparable capabilities, and now have a distribution of attack units which is as wide as that of the insurgents in Iraq. Furthermore, the fact that both Iraq and Colombia currently have the same power-law slope as non-G7 terrorism suggests that the attack unit structures in all three arenas are currently the same.

This is all quite remarkable. Nobody would disagree that wars are terrible human tragedies, charged with emotions and irrationality. And yet it looks like they can be interpreted – and even understood – using Complexity Science.

But all this rests on the quality of the data. So what if someone has artificially inflated or deflated the casualty numbers for Iraq or Colombia? After all, there is likely to be a great temptation to do so depending on which side is reporting the numbers. Well, it is a particular feature of Mike and Jorge's work that their analysis tends to avoid such problems. It turns out that the slope of a power-law is insensitive to the total number of casualties. Any systematic multiplication of the raw numbers by some constant factor has no affect on the slope and hence the value of α. This is because a power-law looks at the *pattern* of events and hence casualties, as opposed to simply monitoring the aggregate number. For this reason, Mike and Jorge have done something which is far more subtle and insightful than merely tracking how the total number of casualties changes over the lifetime of any given war.

Let's think about this for a moment. Imagine someone has given you the total number of casualties in two insurgent wars, and that these numbers are similar. It doesn't actually tell you very much since these numbers might be explained by the fact that the countries involved were of a similar size, or by the fact that the insurgent armies involved had a similar total size. The far more important question is whether the *way* in which the two insurgent groups are waging war is similar. This is what one would like to detect – and this is precisely what the power-law analysis of Mike Spagat and Jorge Restrepo has detected. In short, they have used a Complex Systems focus to uncover a common hidden character in modern wars – and that is something very significant. Furthermore, with the help of Ben Burnett and Alex Dixon, they are now extending their mathematical model to describe the co-existence of several different "species" of insurgent groups – in other words, a true ecology of conflict.

9.5 The timing of attacks

The work of Mike Spagat and Jorge Restrepo tells us an enormous amount about the character of wars. However, as anyone on the ground knows, it would be even more useful to know something about the pattern of attacks in time – in particular on a daily scale. Judging from the news from Iraq that we hear, it certainly doesn't seem like there is any pattern. Monday there might be two attacks in Baghdad, with five casualties in one attack and thirty in the other. Tuesday, there might be one attack in Basra with ten casualties. And so it goes on. So is there any method at all underlying this madness?

It turns out that there is. Sean Gourley and Juan Camilo Bohorquez, working with Mike Spagat and Jorge Restrepo, took the output time-series from this Complex System – in particular, they took the list of the number of attacks per day in Iraq – and started looking for patterns. Unfortunately most statistical tests require lots of data – and the Iraq war is a one-off event, so it only has one set of data. The researchers were therefore faced with a problem which is analogous to the following situation. Imagine someone has told you that they have shuffled a deck of cards. You don't believe them, and so you want to check. If they have indeed shuffled them, then the sequence in which the cards appear should look random. But what does this mean? It means that the actual sequence of cards should look similar to a deck which has been thoroughly shuffled. Now let's suppose that the sequence of cards in the deck represents the sequence of attacks-per-day in the Iraq war. In particular, each card represents a day, and the number of points on each card represents the number of attacks on that day. For example, the three of clubs, hearts, diamonds or spades would correspond to a day with three attacks. Hence the total number of attacks that the insurgent force can produce over the length of the war, is equal to the total number of points in the deck. What Sean and Juan Camilo wanted to find out is if there is any specific order in which the insurgent force is performing these daily attacks – in other words, if there is any specific order in which the cards are arranged?

The card analogy gave Sean and Juan Camilo the clue as to how to proceed with the real Iraq data. They took the deck of cards – or equivalently the set of attacks-per-day – and shuffled them thoroughly. In doing so, they produced a "random Iraq war" in which the numbers of attacks on consecutive days are unrelated. They then repeated this process in order to obtain a large set of such random Iraq wars. Since this analysis of the number of events doesn't involve the size of each event, each of these random wars has exactly the same distribution of casualties as the actual Iraq war, i.e. it would produce exactly the same power-law as in figure 9.3 and with the same slope. However, the order in which the attacks-per-day occurred would be different in each version. By repeating this procedure many times, they were able to get a picture of what the war in Iraq would be like if the sequence of daily attacks was random.

What they have found so far is remarkable. The actual sequence of daily attacks in Iraq shows more order than for a random war. In other words, there does indeed seem to be some systematic timing in the attacks and hence some forward planning by the insurgent groups – just as we would expect from a Complex System containing a collection of competitive, decision-making agents. Going further, they have been able to deduce the particular sequences of daily attacks which occur more often than expected, and those which occur less often than expected. What is even more surprising is that they find similar results for the case of Colombia. Needless to say, they are currently hard at work on further tests to uncover the full extent of the temporal patterns underlying such attacks.

Chapter 10

Catching a cold, avoiding super-flu and curing cancer

10.1 Natural-born killers

Diseases represent a particularly dark side of Complexity. In particular, the most lethal diseases have managed to tap into the heart of what makes a Complex System so difficult to predict, manage and control – thereby outsmarting the body's sophisticated, but ultimately limited, defense mechanisms. Cancer is a particularly powerful example. There may be many others, either lurking on the horizon or yet to be created.

While we as a Society are focused on fighting off old and new threats to our health, so too the world of the pathogen is also becoming more complex. Indeed, the natural process of evolution is continually working against us by allowing pathogens to mutate – with the possibility that any one of these new forms may have the ability to leapfrog over our body's defenses, or resist our man-made medications.

Collections of potentially lethal pathogens, including viruses and bacteria, are continually interacting with us and our immune systems. In short, we are surrounded by natural-born killers. At the time of writing this book, avian or "bird" flu is on the increase. Indeed, most scientists believe that a global epidemic of deadly flu will strike the human population soon, and bird flu is the most likely trigger. In particular, there is one virus, H5N1, that causes avian flu and which is particularly worrying. Most experts believe that the virus will soon adapt itself in such a way that it can spread easily among humans – probably in very much the same way that human flu and the common cold are spread.

Human flu shouldn't be taken lightly either. In the U.S. the human flu season typically occurs between October and April and leads to about 10 percent of the population contracting the disease each year – in other words, millions of people. In addition some 30,000 people die from its complications each year. What makes influenza so difficult to prevent is that the viruses are always changing. Our immune systems can adapt gradually by producing new antibodies after exposure to a virus; however, if the virus mutates very quickly or dramatically most people's bodies are effectively defenseless.

Even the common cold can prove more than a match for us. According to the major U.S. websites which deal with common-cold issues, the average American adult gets about three colds every year and the annual total for the U.S. is about 500 million colds. The common cold is the most frequently acquired illness in the U.S. and causes the loss of more than 100 million workdays and 20 million school days each year. It therefore costs the U.S. about $50 billion each year. This makes the common cold far more expensive than diseases such as asthma or even heart failure. Furthermore, the common cold can result from more than 100 different types of virus, each of which may have several strains. And the only reward for having been infected with one of these viruses is that we are briefly immune to re-infection by that same strain – but just that strain. As a result we can actually catch "a cold" over and over again, as any long-suffering parent of school-age children knows.

10.2 From communities to classes

It has been estimated by U.S. government researchers that a super-flu such as a human transmitted bird-flu would spread fastest among children of school age. It is thought that it would infect about 40 percent of them, and that this number would then decline with age. The overall health costs – neglecting the cost of disruption of the economy – are estimated to be at least $180 billion for even a moderately bad outbreak. All such estimates, and hence contingency plans, are based on our current understanding of how the virus will spread around the community.

But how *do* such influenza-like viruses pass around a particular community? Most theories of how transmissible diseases pass among a population focus on treating the population as a large homogenous group in which everyone is treated equally. In particular, everyone has the same chance of receiving or passing on the virus. But clearly this cannot be right. If someone lives on a desert island with no outside contact they will be far less likely to either get or transmit a particular virus which happened to be at epidemic levels elsewhere on the planet. By contrast, a teacher in a school would have a far greater chance of picking up the virus.

The world's population consists of a collection of objects – people – who are connected together in different ways. Many people have very strong connections with a certain subset of other people – for example, a teacher in a class of children. In other words, we tend to be organized into communities of one form or another and this will tend to dictate our chances of picking up and transmitting a virus. If our own particular community is isolated from any infected communities then the chances of someone in our community picking up the virus are relatively low. So the community structure – or network as discussed in chapter 5 – is very important in determining how a virus passes. Given the fact that we are naturally organized into towns and countries of varying sizes and with varying levels of inter-connectedness, the way in which a potential killer virus such as bird-flu will pass across the planet is not at all obvious.

So policy makers face a fundamental problem regarding what to do in order to reduce the spread of a given disease. In particular, imagine that an unknown virus, or viruses, appears. What should be done at the level of each community in order to reduce the chances of it spreading? More specifically, given that public resources are always finite, how much effort should we spend on controlling the transmission *within* communities as opposed to the transmission *between* communities?

This is the question that Roberto Zarama and Juan Pablo Calderon at the Universidad de Los Andes in Bogota, Colombia, asked themselves. In particular, they wanted to carry out an experiment in which they could study the transmission of viruses within a population which was arranged into communities, and yet where there was contact between these communities. Without having to know the particular virus or its strains, they wondered if they could then deduce something about the effective connectivities between people within a given community, and then between communities, and therefore be able to say something about possible strategies to reduce or contain the virus. In addition, given the possible virulence of flu-like viruses among children, they wanted a large part of the study to focus upon children. For this reason they hit upon the idea of studying the common cold in a large school containing many classes.

10.3 Kids, colds and contagion

There have been many studies of the common cold, although most of these have been in a single community such as a prison or a submarine. By contrast, what seems to be important in real-world situations involving flu-like viruses and superbugs is the competition between transmission within a community and transmission between communities.

Most social systems involve clusters or "communities" within which people have many interactions. Interactions between communities tend to be less strong, or less frequent, but nonetheless do still exist. Unfortunately, most theoretical models of disease transmission tend to ignore such community structure since it makes the analysis too complicated. Roberto and his colleagues came up with the idea of carrying out the experiment in a school since a school contains natural communities in the form of classes. In other words, children interact with other children in their own class – what we could call intra-class interactions – and then at recess will interact with children in different classes – what we could call inter-class interactions. These interactions will also involve teachers, and the teachers themselves interact at recess.

So they went ahead and contacted the school, Colegio Nueva Granada, a U.S.-run school high up in the Andes in Bogota, Colombia. The principals, administration and teachers – in particular Dr. Barry McCombs, Dr. Barry Gilman, Ms. Natalia Hernandez, and science teacher Anne Gregory – were all extremely supportive and helped organize the infrastructure for this school-wide project. There are several important factors that make the resulting study that they did unique in scientific terms:

(1) The Colegio Nueva Granada is one of the largest overseas U.S.-run schools in the world, with a population of close to two thousand and with kids from the ages of four to eighteen. Hence it covers a wide age-range. It is also a relatively tight-knit community, which means that siblings tend to attend the same school, and kids and parents tend to mix with each other out of school hours. This means that it is also a relatively isolated system in that anyone who catches a cold during the school year will probably have caught it from someone else in the school. This is not always the case, of course, but it should happen fairly frequently.

(2) Since the school sits close to the equator, it effectively experiences no seasons. Hence the researchers could, to a good first approximation, ignore the possible effects of seasonality which tend to plague such epidemiology studies elsewhere.

(3) The younger kids are organized into classes, and hence would be expected to interact more frequently with classmates than with children in other classes. By contrast, the older kids are more strongly mixed because of the lack of such a rigid classroom structure higher up in the school. There is therefore a natural community structure – yet just like the real world, it is non-trivial in that the communities are not all of the same type or size, nor do they have the same degree of connectedness (i.e. connectivity).

(4) Since a cold typically lasts for about one week, the process of measuring who has a cold only needs to be carried out once a week. In other words, the kids can be asked once a week whether they have a cold or not, and this should then be often enough to track the spread of the cold. Now it might seem that this is a minor detail – but it isn't. Teachers are busy people and typically have a million and one other things that they have to deal with in any

given week. Given the fact that the researchers could not afford to have any missing weeks in the data collected, this feature was crucial for ensuring that the study could be implemented successfully over a long period of time.

(5) The study could be easily extended to look at the spread of colds within, and between, sets of schools. Hence the study had the scope to span a wide range of community sizes – from people interacting within a class, to classes interacting with each other, to year-groups in a given school interacting with each other, to schools interacting with each other within a given city, to schools in different cities and even countries interacting with each other. In addition, the study requires no special equipment – instead it is possible to apply the same methodology to any school anywhere in the world and compare the results. In fact, this is what the researchers are now working on. Their goal is to have all the data available on the Internet and updated weekly, thereby building the largest ever study of viral transmission in history.

So every Wednesday morning in the Colegio Nueva Granada each student is asked by their teacher if they have a cold. The teachers and students then help put this information in a database ready for analysis. Similar information is also passed into the database about the teachers, administrative staff, canteen staff, and even the bus drivers. With the help of Ana Maria Fernandez at the nearby Universidad de Los Andes, this database is then analyzed. Every person is assigned a unique but secret barcode to identify them, and a 0 or 1 is placed next to their name for that particular week: the 0 means "does not have a cold" while the 1 means "does have a cold". This data then has to be cleaned up since there may have been several students absent. In particular, there is a check the following week whether the person had been away from school with a cold or not. Students who report having colds over several weeks are also checked – just to make sure that they are reporting a genuine cold and not some kind of ongoing sinusitis. Each week, the database grows by adding a further column of 1's and 0's with approximately 2000 entries corresponding to students, teachers and staff. Putting all the weeks of data together then leads to a mathematical picture of the evolution through time of colds in the school.

But how would you analyze such a large amount of data? The data added each week consists of approximately two thousand 1's and 0's – mostly 0's since only about ten percent of the school's population have a cold in a given week, and hence there are only ten percent of 1's. Moreover, this then changes from week to week over the course of the school year. Like so much in science – and in particular, Complexity Science – there is no right answer or approach in terms of the analysis of this data. Basically you just have to try something sensible. The team actually tried two things. First they used the idea of networks which we looked at in chapter 5. In particular, they analyzed the data collected from each set of two consecutive weeks in order to work out what the network of transmission of colds might have been during that week. In other words, they drew out on a network all the possible routes that these colds might have taken from one week to the next. The way they did this was to draw a connection from anyone who had a cold in a given week, to all the people in the following week who had a cold but hadn't had one in the previous week. In other words, a connection going from person A in week 1 to person B in week 2 means that person A had a cold in week 1 while person B did not have a cold in week 1 but did have a cold in week 2. If person B also had a cold in week 1, no connection was drawn since person A was not responsible for B's cold in week 2. Repeating this procedure for the whole school produced a network which is shown in schematic form in figure 10.1.

The data from weeks 1 and 2 produces a single network as shown in figure 10.1. The data from weeks 2 and 3 also produces a single network; however, this will differ from the network corresponding to weeks 1 and 2. In this way, the researchers built up a set of changing networks – essentially a movie of the possible transmission lines of the common cold. They then analyzed it from the perspective of what seemed to be the most active nodes. It may not come as a surprise to long-suffering teachers everywhere, that some of the most active nodes in classrooms containing the younger kids were actually the teachers. In other words, these teachers seemed to act as super-receivers and super-givers of the common cold. They received more colds, and they gave more colds. Poor teachers! Then the research team turned to look

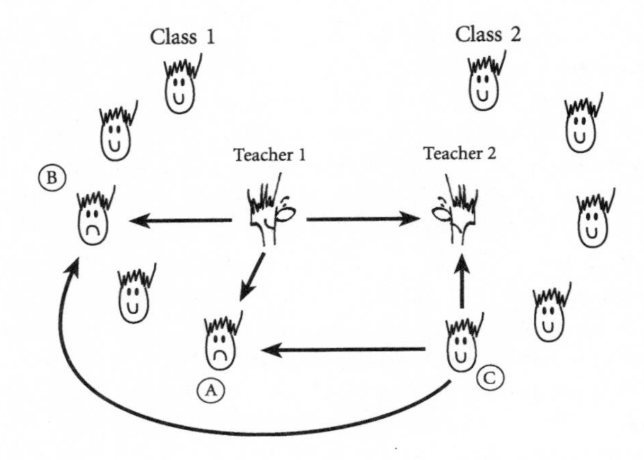

Figure 10.1 Where did that cold come from? Schematic network showing that between weeks 1 and 2, teacher 1 may have given a cold to students A and B in class 1, and to teacher 2 in class 2. Likewise student C in class 2 may have given a cold to teacher 2 in class 2, as well as to students A and B in class 1. This implies that teacher 1 and student C both had colds in week 1 while teacher 2 and students A and B did not. In week 2, neither teacher 1 nor student C had a cold, but teacher 2 and students A and B all had colds.

at the connectivity between people in the same class and between classes. In other words, they tried to deduce the effective connectivity of a typical child in terms of the other children in his or her own class, and then in terms of the children in other classes. This idea is shown in figure 10.2.

With the help of Chiu Fan Lee, Juan Pablo Calderon, and Jameel Kassam, the researchers managed to develop a mathematical model of what was going on. In particular, they found that they could explain the transmission of the colds in the school using a theory in which each person within a given class 1 is considered to have, on average, close contact over a given week with a number of people K_c within his or her class, and a number of people K_b in the remaining classes 2, 3, 4, etc. They then used a very common trick from physics, whereby these remaining classes were assumed to form part of a "sea" or "bath" of students and

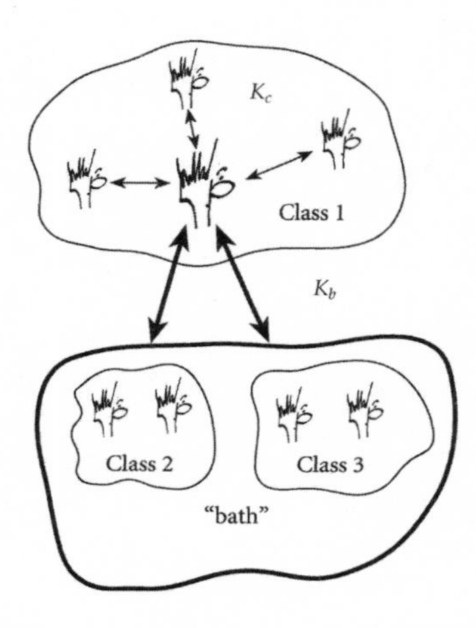

Figure 10.2 It's who you know that counts. Each person in class 1 has, on average, close contact over a given week with a number of people K_c within his or her class. He or she also has close contact over a given week with a number of people K_b in the remaining classes 2, 3, 4, etc. The mathematical theory that the researchers developed was simplified greatly by assuming that these remaining classes 2, 3, 4, etc. formed an effective "bath" or "sea" in which the identity of which class is which, and which person is which, can be largely forgotten. This is a common approach in physics whereby systems of objects get replaced by a background "bath".

teachers. In other words, everyone who was not part of class 1, became part of a bath with which class 1 interacted. Viewed from the point of view of class 1, the bath is a sort of blurred version of the rest of the school in the sense that the identity of which class is which, and which person is which, is forgotten. For a person in class 1, anyone who isn't in class 1 is just part of this big bath. And the same is true when viewed from the perspective of class 2, except that the bath for class 2 now contains class 1 as opposed to class 2. So every class sees every other class as part of a big bath. It is well-known in physics that this type of approach can prove very successful in describing the behavior of quite complicated collections of interacting particles.

Using this mathematical model, the research team were able to deduce the following information from the data. As the average

age of the class being considered increases, the number of people K_c with which a person in that class typically has close contact, decreases. By "close contact", the researchers mean "sufficient contact in order to transmit a virus". This does not mean that the typical person in that class necessarily has K_c close friends – it just means that he or she has sufficient contact with them in order to pass on the virus. This contact may be due to proximity – in short, within sneezing distance. The contact may also be indirect – such as a cold virus left on a pencil which is later used by the other person. By contrast, they also found that as the average age of the class being considered increases, the number of people K_b in the bath with which a person in that class typically has close contact actually increases.

It is amazing that these important numbers K_c and K_b can be deduced from this real data involving kids, colds and classes. And the results make sense – as the kids get older, they spend less time interacting in classes (hence K_c decreases) and more time in common activities involving all the kids in their year-group (hence K_b increases). An additional reason why K_c would be expected to be higher for younger children is that they tend to have closer physical contact with their peers, to be less careful to cover their mouths when they sneeze, and to be generally less conscious about personal hygiene – as any parent will again know too well.

Interestingly, this same model of transmission of viruses in communities can be applied to the conflict situations of chapter 9 in which violence spreads like a virus between neighboring communities – and where violence in one community can spark violent acts in another. It can also be used to describe the transmission and hence spread of news or gossip in financial markets. Each community is now a particular market, or market sector, or even a particular currency as in the foreign exchange markets mentioned in chapter 5. Just like the colds, some rumors stay within a given community while others jump between communities if the number of contacts is large enough. This colds-based project therefore provides us with valuable insight into how objects such as a virus, news, or rumor, get transmitted through a complex network of connected groups of people. In particular, it teaches us that it is the *differences* in the connectivity within and between communities that can play a dominant role in determining the pattern of transmission.

The implications for controlling the spreading of viruses and rumors are also interesting. Without going into too much detail, the results of the study suggest that if a super-virus is spreading through a population and is, as has been suggested, hitting children particularly hard, then the approach for control should be as follows. For younger children, attention should be focused on reducing the amount of contact within their own class – for example, by separating them more within the classroom. This will lead to a lower number of contacts K_c within the class – and since K_c is typically larger than K_b for younger children, this should help reduce the spread. For older age-groups the situation becomes reversed: less attention needs to be placed on the contacts K_c within a child's class, and more on the contacts K_b between classes.

The research team are currently extending this study to other schools – and treating this extension within the mathematical model as yet another layer of communities. They are also developing the multi-community transmission model to understand the spread of news and rumors through the various currencies and stocks in a financial market, and between different global markets. In related work, Elvira Maria Restrepo at the Universidad de Los Andes is applying a similar idea to analyze the spread of crime within and between different districts of large cities.

10.4 Cancer: how to starve a tumor?

Unfortunately, viruses and rumors aren't the only dangerous things that spread. Cancer is something which kills by spreading. After the initial tumor has taken hold, it may spread out and become so large that it engulfs the organ to which it is attached. What's worse is that even though such primary tumors can often be removed, cells from it may have already started spreading to other parts of the body, creating a variety of secondary tumors which are not detected until it is too late.

Cancer is a terrible and tragic example of a Complex System. In fact, it shows Complexity on all levels: from the set of microscopic and genetic processes that need to go wrong in order for it to start, right through to the way in which it manages to trick the body into supplying it with enough nutrients to grow and flourish – even though this may mean that the body itself could die. But exactly how it does all this is a mystery. As a result of the recent expertise in genetics which has arisen in the medical research community, there is a lot of effort being spent on trying to understand the first aspect of this Complexity – in other words, looking at what might be going on at the genetic and molecular level in order that the body then sets off on a course for cancer. The underlying hope which drives all this research is that cancer might conceivably be nipped in the bud before it fully develops.

However, the massive problem facing cancer researchers is that there are many reasons why a given cancer might start. This is why the media seem to be forever announcing that scientists have found a gene, protein or molecular process which could be important in the development of cancer. Unfortunately it is likely that even if we heard such announcements every day for the rest of our lives, scientists would still not have uncovered the full range of possibilities for how cancer can be produced. The point is that cancer tends to start like an error at the microscopic gene-protein level, and these errors can arise in presumably an infinite number of ways. It is like asking how many spelling mistakes can be made by someone copying a piece of text, where in this case the text corresponds to the DNA code or the molecular-level instructions for some protein production process. The chance of us ever being able to classify all possible errors is minimal – in fact, it seems to be true that among all the cancers ever studied, no exact same error has ever been observed.

Less attention tends to have been paid to the second aspect of Complexity mentioned above – the supply of nutrients which any given cancer, like a plant or fungus, needs in order to survive and grow. Yet this is precisely the area where the ideas of Complex Systems and networks, as described in chapters 4 and 5, might prove useful. This idea is being pursued by Sehyo Charley Choe, Alexandra Olaya Castro, Chiu Fan Lee, Philip Maini, Tomas

Alarcon and others – and it makes sense for the following reason. It is quite likely that many of us have very small, embryonic tumors already sitting inside us – yet they may never become life-threatening. Such small tumors are typically unable to grow any larger because they do not have a nutrient supply nearby. The nutrients for a tumor come in the form of oxygen and glucose which are carried in the blood. Hence blood vessels are needed to supply the tumor with these nutrients and to carry away any waste products. As an analogy, just think of a vehicular transport system by means of which food is supplied to a given town (i.e. tumor) and trash is carried away. If there is no road, there are no vehicles – and if there are no vehicles then there is no food and hence the town will not grow. Indeed, it seems that we can all survive quite happily with such small tumors and will remain oblivious to the fact that they even exist.

Unfortunately for us though, this situation of small but essentially static tumors may not last forever. At any particular time, the tumor may suddenly trigger the growth of blood vessels toward it. This is very much like our starved town managing to convince the roads to grow themselves toward it, and hence deliver nutrients as shown in figure 10.3. The tumor – or more precisely, the cancer cells within the tumor – perform this trick by releasing chemicals

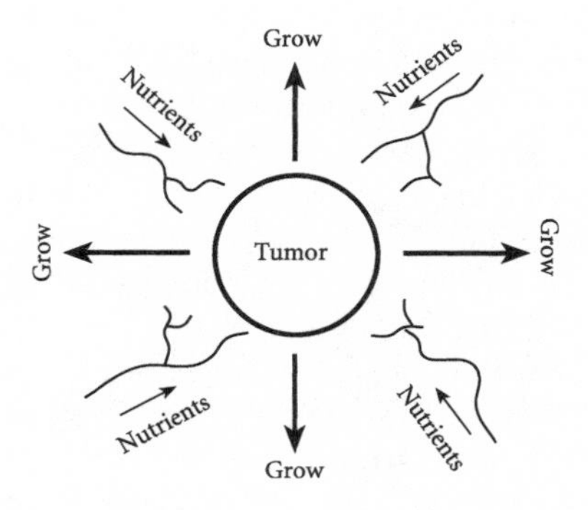

Figure 10.3 Eating nutrients leads to growth. The nutrients are supplied through blood, which in turn flows along the "roads" formed by blood vessels and capillaries. The road network itself is called the "vasculature".

which promote the growth of new blood vessels or capillaries near the tumor. This process is called angiogenesis. One of the first people to show the importance of this process was Dr. Judah Folkman of Harvard University. The treatment which then suggests itself is to use drugs which reduce the effects of angiogenesis. This approach could run into trouble, however, since angiogenesis is also a process that helps heal wounds. The patient might then be in danger of dying from wounds that don't heal. Indeed, this is why the skin around a cut turns red – blood vessels are being grown in order to help heal the wound.

As a result of angiogenesis, a growing tumor will tend to modify its underlying nutrient network, thereby giving rise to very strong feedback between the structure of the nutrient network, the size and shape of the tumor, its function and hence its lethality. And just as the efficiency of a transport network can be affected by the fact that short cuts may or may not exist – in other words, by the way in which it is wired – a tumor's growth can be affected by the way in which the capillaries are arranged. In particular, going back to the discussion in chapter 7, we know that it is possible for two networks to have very different structures and yet have the same functional properties in terms of the average shortest path from one side to another. The same holds for cancer tumors – there can be arrangements of blood vessels and capillaries which are structurally very different, and yet which are equally efficient (and hence equally lethal) in transporting nutrients to all parts of the tumor. In other words, two blood-vessel – or so-called "vasculature" – networks in two different tumors may look very different, but they may have exactly the same functional properties and hence the same level of lethality. This is one of the reasons why it is so hard for doctors to predict how lethal a particular tumor will be.

The connection to topics discussed earlier in the book doesn't stop there. It turns out that there is a war going on within the tumor at a more microscopic level. In particular at any given place within the cancer tumor itself, there are two competing populations of cells: cancer cells and normal cells. Cancer cells will do just about anything in order to survive. Not only do they fight to receive nutrients with which to grow, but they also fight for space into which they can grow. Indeed, this sounds reminiscent of the competition for space in the bar problem in chapter 4, and the wars in chapter 9. More generally it is reminiscent of the competition for limited resources which this book has flagged as a fundamental feature of real-world Complex Systems. Indeed, the fact that cancer cells have such complicated interactions which can change in time means that we can usefully think of them as having strategies. Unlike normal cells, cancer cells evolve in such a way that regulatory mechanisms are avoided and they therefore represent a population of competitive individuals – very much like our agents or traders in a financial market, or drivers in traffic. And as we have seen throughout the book, a collection of such agents competing for limited resources can act in complicated ways. Hence it seems like the process of tumor growth should embody all the main features of Complex Systems discussed in this book.

Most previous mathematical models of cancer tumor growth have relied on treating this competition and nutrient-supply problem in an average way. In other words, just like the use of the bath in the colds project earlier in this chapter, the tumor is treated as a rather blurry object with no structure. However, for a tumor the devil is definitely in the details of how and where the nutrients are being supplied – so such an approximation will be unreliable in general. By contrast, Sehyo Charley Choe and Alexandra Olaya Castro have been developing a new model which combines all the elements of Complex Systems which we have discussed so far. In particular, it features agents (i.e. cells) fighting it out on the microscopic scale for space and nutrients. Their actions then feed back onto the underlying nutrient network which in turn feeds back onto the agents themselves. The idea is to use this model to see what changes in vasculature need to be made in order to stop a given tumor from growing. After all, if the tumor doesn't grow and all activity ceases within it the patient is safe for an indefinite period of time – most importantly, that cancer won't kill them.

Charley and Alexandra's model allows us to examine the types of tumors (i.e. size, shape and growth-rate) which can emerge from the interplay between the competition among cancer and normal cell "agents", and the underlying vasculature nutrient network which is necessary for tumor growth. The case of skin cancers is particularly interesting – not only because it is one of the fastest growing forms of cancer in the population, but also since these tumors are usually more rugged in shape. It is hoped that an understanding of the interplay between the tumor's functional and structural properties, and the underlying vasculature network, could help doctors in their important yet very difficult task of visually identifying possible skin cancers.

So how can we beat cancer? Charley and Alexandra's model shows that if we can restrict the tumor's nutrient supply by reducing the underlying vasculature, we may not only be able to stop the tumor from growing but may also help shrink it. They are now analyzing how different initial vasculature patterns may either favor or inhibit a given tumor's growth, and how doctors might therefore be able to manipulate or re-wire this underlying vasculature network in order to effectively kill off the tumor. A sort of "tailor-made starvation". This work is on-going, and has the potential to produce some very valuable diagnostic tools. But to help us understand how it might work, let's just think back to chapter 7, where we saw the effects of adding a cost to a hub-and-spoke network. Thinking of the hub as the tumor (see figure 10.4), the implication of chapter 7 is that if a "cost" can be introduced for transporting nutrients to the tumor then the number of

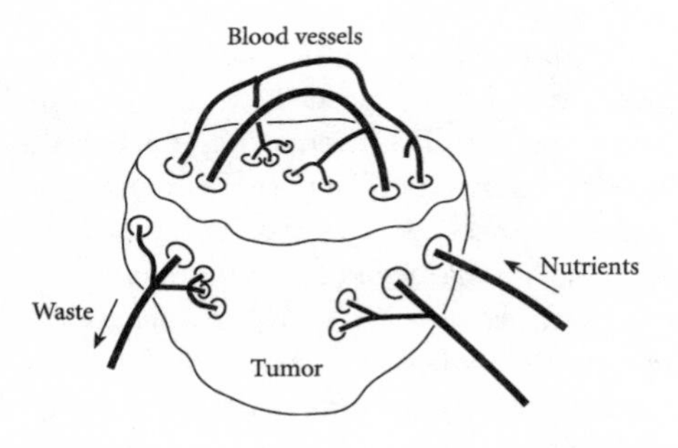

Figure 10.4 Starving the tumor. If the flow of traffic (i.e. nutrients) through the tumor could be reduced, the tumor could be starved and hence would stop growing.

connections and hence nutrient pathways into the tumor can be kept small. More specifically, if the cost of connections to the hub can be made high, no new blood supply highways (i.e. major blood vessels which provide an influx of nutrients while simultaneously carrying off waste) would arise, and any existing ones would die out. In short, the tumor would become benign.

10.5 Showdown: Superbugs vs. the immune system

Our body's immune system is our first and last line of defense against all sorts of bugs (i.e. pathogens) that have the potential to make us seriously ill, and even kill us. Sounds important – and yet the mechanisms of the immune system are very complex and poorly understood, even by immunologists. This is because the immune system is actually a collection of objects which are themselves complicated – and the whole thing is held together by a remarkably intricate web of interactions and feedback processes. It is therefore a typical Complex System.

In computer systems and information systems, software equivalents of the immune system are being developed and employed on a daily basis in order to prevent contagion from the multitude of viruses being created. Our own bodies benefit from an immune system that can adapt itself to new challenges – or rather, can do so up to a point. However in computer systems, the software needs to be written each time a new virus appears. Hence there is an enormous amount of interest, both academic and commercial, in learning relevant tricks from our biological immune systems – in particular, how to build adaptability into software systems whose job it is to protect the system against present and future attacks by old and new viruses. We only have to look back to chapter 9 and the tales of conflict to appreciate that there is also an important parallel with asymmetric wars whereby a small insurgent force might attack a large body with potentially lethal consequences. Interest in how to build a suitably adaptive and protective defense therefore also arises in the military domain. On a lighter note, we can also think of the dating example in chapter 8 whereby we – presumably in the style of Victorian-age parents – might wish to introduce a chaperone as "defense" to stop interested males/females from approaching our supposedly defenseless daughters/sons.

So what *is* the best form of defense? In particular, if a system such as the body, or a computer node, or a city, or even a country, is being attacked by some unknown superbug, how should we organize our defense? Is the best form of defense to go out in search of the attacking force – hunting it down? Or is it to encircle the object we are trying to defend and just lie in wait for any attackers? Both methods have their advantages and disadvantages. But how can we model this?

At the end of chapter 8 we mentioned this exact same problem, albeit in another context. In particular, Roberto Zarama, Juan Camilo Bohorquez and Tim Jarrett have created a sheep-dogs-wolves game to mimic an attack by unknown predators (i.e. the wolves) on a set of defenseless objects (i.e. the sheep). More specifically, they have modeled a scenario where the attacking agents (i.e. the wolves) wish to destroy the target agents (i.e. the sheep) and hence win the game. The targets are protected by defensive agents (i.e. the dogs) which we assume that we can control, at least partially. The complications arise from the fact that we don't know how many wolves there are, nor where they are – nor do we know where the sheep are. And nor do the dogs. So do we send out the dogs to look for wolves, in the hope that the dogs can find the wolves before the wolves find the sheep? The problem with such an attacking defense strategy is that if the wolves outnumber the dogs significantly, and each dog ends up chasing a wolf, then there will be an excess number of free wolves who may simply wander up to the sheep and kill them. On the other hand, having the dogs try to track down and encircle the sheep from the outset may also cause problems – the sheep will generally change their position in time and hence the dogs would need to constantly track the sheep better than the wolves can. One false move by the dogs and the sheep are dead. Indeed, even the problem of one sheep, one dog and one wolf will be complex – it is a three-body problem, and when combined with motion on a complex network will give rise to all sorts of rich behavior.

Tim Jarrett has studied what happens in this game when the objects move on a complex network, such as the hub-and-spoke of chapter 7. He allows the wolves and dogs to smell other nearby animals in order to sense their direction, which effectively gives the wolves and dogs hunting abilities. He then explores two types of behavioral model for the defending objects (i.e. dogs): attacking and defensive. We will refer to these as attacking defense and defensive defense, respectively. In the case of attacking defense, the dog attempts to defend the target by attacking the attacker. In the case of defensive defense, the dog attempts to defend the target by sitting next to it. Overall Tim finds that for small numbers of dogs, sheep and wolves, *attack is the best form of defense.*

This model can be extended in all sorts of interesting ways, and to many other applications. For example, one can imagine that the dogs play the role of a police force, the sheep are ordinary citizens, and the wolves are criminals. How does one then deploy a limited number of police in order to best protect the community? In the case of just a few criminals, Tim's results suggest that the strategy of tracking down known criminals even before they have committed the next crime, would be a useful form of defense – as opposed to trying to defend directly a cross-section of the population. He is currently looking at the effects of either losing a defender, adding an attacker or corrupting a defender in order to measure the system's robustness. For example if one allows for the possible corruption of a dog into a wolf, the best policy may be to send out dogs in a crowd of three or more.

It turns out that a very similar problem arose for the allied Navy during World War II, but in a completely different context. When convoys of ships were being sent across the Atlantic in order to supply the allies and their forces with food and machinery, the Navy had to decide how to defend these ships. These supply ships typically had no defenses of their own, and hence represent the sheep in our game. The accompanying frigates and battleships played the role of the dogs, and the role of the wolves was very vividly played by the German U-boat submarines. The Navy did not in general know where the wolves were, nor did they know how many of them there were – and yet the convoy needed to have some kind of strategy. In the end, the Navy decided to introduce a black-box device on the ships which, every so often, would spit out instructions for random changes in course. The result became very much like a herd of sheep who start wandering semi-aimlessly across a field – or in this case, the ocean. The German U-boats would then have had a hard job tracking the boats in order to torpedo them. Complexity and conflict – hand in hand once again.

Chapter 11

The Mother of all Complexities: Our nanoscale quantum world

11.1 Einstein's spookiness

Nobody would deny that Einstein was a genius. But it turns out that he was a slightly tortured genius – tortured by something for which he had earlier been awarded the Nobel Prize. That something is *Quantum Physics*. It turns out that although we know of Einstein most typically through his work on space-time and relativity, he was actually awarded the Nobel Prize for his work on explaining the so-called photoelectric effect. Einstein's great insight was to explain this effect – whereby particles called electrons are emitted from a metal using light – by means of quantum physics. In so doing, he revolutionized the understanding of our world.

The story of this discovery is as follows. People had shown that if light is shone onto a piece of metal, the light can kick particles called electrons out of it. Maybe that doesn't seem so surprising? After all, light has energy. And surely the brighter the light, the more energy it has, and the more particles it will kick out. After all, if I shake something like a tree harder and harder, more leaves are going to drop off. But this is not what was observed.

At first scientists had tried using a very dim red light, but it did not liberate any particles. One would therefore imagine that sending in light of any other color but with the same low brightness, would also have no effect. That is not what happened. In particular, they found that dim blue light did liberate particles even though dim red light did not. So the energy of light has something to do with color. And here comes the real surprise – turning up the brightness of the red light has absolutely no effect at all. Even if the most intense red light you can imagine is falling on the metal, no particles come off. And yet a very dim blue light *is* enough to release particles. So what is going on?

The explanation that Einstein gave is that instead of thinking of the effect of light as a shaking motion (or in technical jargon, a wave), whereby the brighter the light the more vigorous the shaking, it is instead more akin to a stream of objects like balls. So throwing light on a metal is like throwing balls at a coconut-shy in a fairground. The light is the stream of balls – and the electrons in the metal that you are trying to kick out, are the coconuts. The red light then behaves like very lightweight balls – balls that are too light to dislodge a coconut even if you throw twenty in rapid succession. In other words, it doesn't matter how many of those lightweight balls you throw at the coconut, the coconut won't move. The same happens with the light on the metal. Red light is made up of lightweight balls, or rather packets or "quanta" of energy – a stream of them acts like a stream of lightweight bullets. Hence if we fire a steady stream of these red light quanta at the metal, none of them is individually energetic enough to dislodge the particles inside. It doesn't matter how rapidly we send them, they are never going to do the job. By contrast, blue light is made up of balls which are relatively heavyweight in terms of their energy. In short, each individual quantum of blue light carries a lot of energy. So even if we fire only one at the metal, it can dislodge a particle – just as we would be able to dislodge the coconut with a single accurate throw of a heavyweight ball.

This effect of light appearing as packets or "quanta" of energy is a direct manifestation of Quantum Physics. In fact everything in our world exists as quantum objects like the balls of light – and we give all of them the generic name of quantum particles. This is all very surprising, and explaining this won Einstein the Nobel Prize. But that is where his problems began. It turns out that although an individual quantum particle is clearly quite a strange object, two or more are positively weird. In fact Einstein labelled the special quantum properties of a collection of two quantum particles, as "spooky". He was right to call them spooky, as we will show in the next section. However Einstein never accepted this spookiness and effectively spent much of the rest of his life trying to show that Nature couldn't possibly be that spooky. And because technology at the time couldn't test out the experiments that he had thought up to challenge Quantum Physics, they just remained as arguments in his brain – or so-called thought experiments. Since these thought experiments couldn't be carried out in practice, the arguments became philosophical and hence were never resolved while he was alive. And like all unresolved things, this must have left Einstein feeling very uneasy.

Fortunately, recent advances in optical technology have enabled Einstein's thought experiments to be performed in a laboratory. However, unfortunately for Einstein, these experiments have shown quite conclusively that Quantum Physics is correct and that Einstein was wrong to dismiss its spookiness. In short, Nature is indeed very spooky.

11.2 Three is a crowd, but so is two

Quantum Physics is truly the "mother of all complexities". It underlies everything in our Universe. Everything from human cells to people, candy bars to airplanes, cell phones to baseball bats. All are made up entirely of quantum particles. We don't notice it in our everyday lives, but it is true. But the real hard-core aspect of quantum complexity emerges when we consider an isolated group of quantum particles – for example, two quanta of light which we can think of like a pair of gloves. The analogy actually makes a lot of sense, since it turns out that a quantum of light has a handedness just like gloves. In other words, a quantum of light is either left-handed or right-handed. We call this handedness polarization, and it is because of such polarization that Polaroid sunglasses are so good at reducing glare. Reflected light from water on the road is strongly polarized in one direction, and so Polaroid sunglasses can be used to cut out that particular direction of polarization while leaving the rest untouched. Hence we see the things around us, but we don't see the glare.

Let's imagine that it is a cold morning, and so you take your gloves to work. At the end of the day, we will imagine that you accidentally leave one of your gloves in the office. When you get home, there is a message on your answering machine from the receptionist saying that she has found one of your gloves. You rush to your bag, and take out the glove that you still have. It is a right-handed glove. Hence you immediately know that the glove that the receptionist has is a left-handed glove. Now, until you looked in your bag, you didn't know whether you had lost your right-handed glove or your left-handed glove. However, this isn't because the identity of the lost glove was some kind of hidden secret that Nature was keeping all to itself. It just reflects the fact that you yourself didn't yet know. But the gloves knew. In other words, anyone could have found out the answer ahead of you – the information was not some kind of fundamental secret of the Universe. The fact that you yourself didn't know is just because you hadn't yet looked or been told.

Now let us consider the same story with a pair of "quantum gloves" – in other words, a pair of quantum particles such as two quanta of light. No store sells them – yet. But we can do a thought experiment just like Einstein. What would happen is as follows. The story is unchanged in all aspects except that until somebody looks at one of the gloves – or as scientists say, until someone measures whether one of the gloves is right-handed or left-handed – even the gloves don't know which is which. In other words, this *is* a fundamental secret of the Universe. Nobody can know the answer as to which glove is which until somebody, somewhere, has looked at one of the gloves and hence effectively measured its handedness. As soon as somebody has done that, then the story is exactly the same as the previous one. However, until that moment of measurement, each glove is both right and left-handed at the same time. Scientists refer to this strange coexistence of both possibilities as a superposition. They also refer to the magical connection that seems to exist between the two gloves – or quantum particles – as entanglement.

Just to emphasize, we do not mean that each quantum glove is momentarily right-handed and then left-handed. We mean it is right and left-handed simultaneously. It exists in both possible worlds at the same time, like two parallel Universes. For this reason, we are looking at a type of emergent phenomenon, and hence Complexity, which is way beyond anything we have seen so far in this book. It is also very spooky.

So if the receptionist doesn't find the quantum glove, and you never look at the other quantum glove again, the Universe will co-exist in two parallel worlds with each quantum glove being both right and left-handed. It is only when someone checks the handedness of one of the gloves that the Universe then collapses into one outcome or the other. And it does this randomly. Einstein said "God doesn't play dice" to which the reply came "Don't tell God what to do!" But the effect really is bizarre – so Einstein's disbelief reflects his deep thinking rather than any sort of lack of insight.

This complex double-life of things – indeed, everything from ice-creams to bicycles if we look hard enough – is really very strange. And it has a huge number of consequences, many of which have scientists still struggling to understand. For example, computers work by storing 1's and 0's. But a quantum computer would be able to store 1's and 0's at the same time. This has led people to deduce that a quantum computer could run faster than any normal computer ever could – so watch out Intel Corporation. Going further, one could use this double-life to create completely safe secret codes. Instead of passing around a secret password to open up a safe box, one could instead pass around collections of quantum gloves with the right-hand being a 1 and the left-hand being a 0. If anyone tries to read your password, such as the receptionist trying to look at one of the quantum gloves, you can detect this tampering simply by seeing if the other glove has "decided" to be right or left-handed.

To understand better this spooky connection between pairs of quantum particles, let's think again about the two quantum gloves. Imagine the situation whereby the two gloves are living in this funny kind of entangled state with them both being simultaneously right and left-handed, while at the same time being gradually pulled farther and farther apart. As long as no one looks at their handedness, they continue to be entangled. In fact, they could be sent to opposite sides of the planet, or Solar System, or Universe – they will still be entangled in this way just as long as no one looks at their individual handedness. Now imagine that someone on Earth then checks the handedness of one of the gloves and finds that it is right-handed. The other glove immediately becomes left-handed, regardless of how far away it is. This seemingly instantaneous action-at-a-distance is truly, truly spooky.

It turns out that many different types of quantum particles in Nature can be created in such entangled pairs – or even in threes. In addition, one can add in particles to a group to make larger groups of entangled particles. And they all share this same type of entangled information whereby they won't decide whether they are right-handed or left-handed until all the others do. In fact, many scientists believe that such entanglement is even more fundamental to Nature than the particles themselves. After all, entanglement seems to carry the information about what a group of particles can do. And since information is the key to everything, then entanglement is arguably the fundamental object in Nature. Scientists are therefore working extremely hard in order to find out what happens when we add together sets of entangled objects – in other words, the quantum equivalent of a crowd. In particular, how does such a quantum crowd behave? Nobody really knows. But it is certainly an amazing aspect of Complexity. Moreover it underpins everything in our Universe. In short, three or more quantum particles is a very strange quantum crowd and so is two – and that certainly qualifies it as the mother of all complexities.

11.3 The secret nanoscale life of plants, bacteria and brains

To what extent does something as mundane as spinach depend on the spookiness of quantum physics? In short, is spinach spooky? In principle it is, since like everything else it contains quantum particles. But would we ever be able to notice this? In other words, why should we care about Quantum Physics when we are munching a spinach salad? Well, it turns out that Life on Earth does indeed use Quantum Physics in a fundamental way – and there are even some recent suggestions that it exploits many of its spookiest aspects. Plants, and indeed many bacteria, produce food through photosynthesis – and experiments using light have shown that the process is very much like the coconut-shy example that we mentioned earlier. The packets, or quanta, of sunlight hit the leaf, and each quantum of light of the correct color then transfers its energy into the leaf. So this process is using the aspect of Quantum Physics that gave Einstein his Nobel Prize.

So far it might sound strange, but not that spooky. However here comes the twist. The packets of energy which then travel around in the leaf or bacteria are quantum particles called excitons. And Alexandra Olaya Castro and Chiu Fan Lee have recently shown that these excitons should be able to exist in an entangled state, at least for a short period of time. Given that photosynthesis is a remarkably quick and efficient process for converting light energy into chemical energy or "food", it is therefore possible that this entanglement is being utilized at some level by Nature. Indeed, Alexandra's calculations have shown that such entanglement may even have a dual role. It can increase the efficiency of the photosynthetic process and it can be used to reduce a possible overload of energy – a sort of crowd control, governed by the type of entanglement. Alexandra's calculations also suggest that one could exploit this same effect to create a completely new class of novel nanoscale devices, including nanoscale solar cells or energy converters for harvesting light.

For technical reasons, what we say below actually describes more closely the process of photosynthesis in purple bacteria than the one in green leaves. But the additional complications that arise in leaves are not important for our story – hence we will just pretend that the process of light-harvesting in green leaves is identical to that of purple bacteria. As we mentioned above, the packet of sunlight gets absorbed and creates the exciton, which is itself just another type of packet of energy. The exciton then gets passed among a network of ring-like structures of different sizes, as shown in figure 11.1. It eventually ends up on the largest ring. In some way which is still not properly understood, it then gets transferred to the reaction center (RC) and converted into chemical energy, i.e. food.

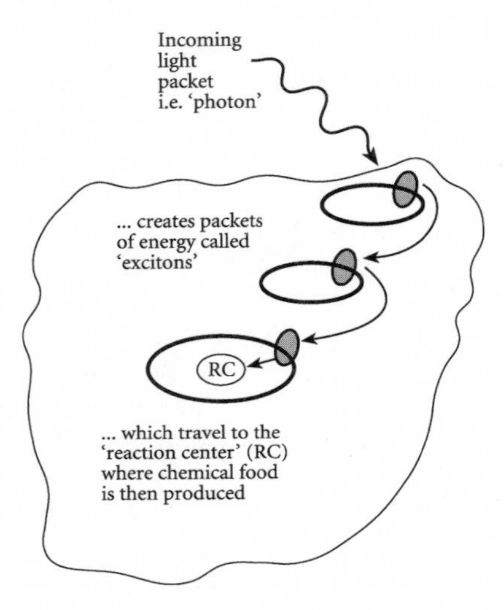

Figure 11.1 The meaning of Life. Sunlight gets converted to food for plants and bacteria. Plants then give rise to food for animals, and plants and animals provide food for us. So without this initial conversion of light energy to food energy in the leaf, there would be no Life for us on Earth. For technical reasons what is sketched more closely describes the process of photosynthesis for purple bacteria rather than for green leaves. But the additional complications in leaves are not important for our story – hence we will just pretend that the two are the same.

Alexandra's calculations suggest that the step involving the transfer of the exciton to the reaction center can benefit greatly from the spooky aspect of Quantum Physics through entanglement. So given that entanglement can indeed help, and given that entanglement is a naturally occurring phenomenon, could it be possible that Nature is already using it? At the time of writing this book, the experiments have not yet been done to test whether entanglement exists in such biological systems. However the rings are made up of proteins and molecules which could in principle be measured for entanglement. The only problem is, *how* would you measure for the presence of entanglement? This is where the work of Ferney Rodriguez and Luis Quiroga of the Universidad de Los Andes in Bogota, comes in. They have produced a general mathematical theory which shows that if one correctly treats the quantum effects in the entangled systems of interest plus the surrounding "bath", the entanglement can be detected using currently available optical equipment and setups. It therefore just remains to be seen what the experimental outcome will be for the photosynthetic rings shown in figure 11.1.

Even more controversial than quantum effects in photosynthesis are quantum effects in the brain. This is an idea that has been recently promoted by Stuart Hameroff of the University of Arizona and Roger Penrose of the University of Oxford. In particular, they have claimed that exotic quantum effects will arise in the microtubules that lie within the cells of our bodies. Microtubules can be thought of as a sort of scaffolding in our cells, giving them structure but also serving a variety of other functions such as providing roadways for transport. Microtubules are made up of a collection of proteins which are arranged in the form of a hollow tube, just like an empty kitchen roll. Stuart Hameroff and Roger Penrose believe that the microtubules in the brain exploit quantum physics and the spookiness of entanglement in order to give rise to brain function and consciousness. In particular, they believe that packets of quanta can survive for sufficiently long in these microtubules such that they end up processing information like a sort of quantum computer. This would be no mean feat, since the quanta in the microtubules – like the quantum gloves – are in constant danger of being "looked at" or measured (and hence losing their entanglement) by all the other chemicals and molecules which sit around in the brain. So far, nobody knows whether the proposal is correct. But it certainly is thought provoking. After all, the brain is arguably the most complex Complex System in our world. So is it not conceivable that it is running on an engine which is powered by the mother of all complexities?

11.4 Quantum gaming

So maybe quantum mechanics is playing a fundamental role in Nature – and maybe we can design artificial devices based on its properties in order to enhance natural processes. But, apart from the possibility of selling such novel devices, can entanglement make us rich? The answer is yes – maybe. It turns out that there is a brand new research field springing up which combines the complexity of games that we saw earlier in the book with the spooky complexity of quantum physics. And this field is referred to as quantum games.

Let us suppose we challenge someone to a coin-flipping game, also supposing that our opponent is a normal human being. We, on the other hand, have quantum superpowers in that we know how to exploit Nature's ability for particles to live in limbo. In particular, we will imagine that we have the ability to generate superpositions and entangled states, such as the glove being right and left-handed at the same time. Now, suppose there is a coin which we prepare in a special way and give to our opponent. Our opponent can either flip the coin or leave it unchanged before handing it back to us. Before the referee looks at the outcome, we are then allowed to flip the coin again if we want to. Our opponent loses if the final outcome is heads, but wins if it is tails.

So the quantum game begins – and we are able to win every time. But how? It turns out we have chosen to use a quantum coin, which – like the gloves – can be both heads and tails at the same time. All our opponent can do is flip the coin or not. However, we can perform a whole range of operations. For example, we can effectively put the coin into a superposition of heads and tails. This means that if our opponent then flips the coin, the heads part becomes tails and the tails part becomes heads. So the coin remains unchanged. After our opponent has completed his move and returned the coin to us, all we need to do is to return the coin to being purely heads in order that we automatically win the game. Guaranteed, 100 percent success.

This idea uses a single object, a quantum coin, and puts it in a superposition of quantum states – heads and tails just like the right-handed and left-handed quantum glove. The idea was introduced by David Meyer of the University of California and really helped to set the quantum games field alight. However in true Complex Systems style, the real power of quantum games comes when we consider slightly more complicated games involving more players. This was first done by Jens Eisert of Imperial College, London and co-workers. Important work has also been done by Adrian Flitney and Derek Abbott of the University of Adelaide in Australia. In addition, a formal theory for such quantum games has been introduced by Chiu Fan Lee. It turns out that quantum games involving three or more players are particularly interesting and complex. Work in this area by Simon Benjamin and Patrick Hayden, and by Roland Kay and co-workers, has shown that there are outcomes in three-player quantum games which are completely different from the standard everyday version of the same game. Our own group has shown that adding a corrupt referee to such quantum games can turn a winning quantum game into a heavily losing one.

Thinking about a possible commercial setup of a quantum game on the Internet, all that would be needed is for each player to submit online his instructions for how to manipulate his own "quantum glove" or quantum coin. This corresponds to an action by each player, and these actions can then be executed on a set of quantum particles which are held centrally by the referee. When all these actions have been executed, the referee then takes a look at (i.e. he measures) this collection of quantum coins – just like a croupier would check the cards in a casino at the end of a set of plays. The referee then announces the winner. Even at this stage of the game, quantum physics can prove its worth by providing a foolproof check on the referee. In particular, sets of quantum gloves can be used to monitor each of the referee's actions.

So there we have it – a completely new type of game where corruption can be detected, and where a huge range of unusual payouts and strategies become possible. Will it catch on? Probably somewhere in the future it will. After all, where there is an opportunity to make money, somebody usually ends up doing it. In more general scientific terms, it turns out that many processes in physics can also be seen in terms of objects playing games. This field of quantum games therefore offers a fundamentally new perspective for helping unravel the complexity and spookiness of Quantum Physics itself.

11.5 Many wrongs can make a right

Let's imagine for a moment that various prototype quantum devices have been made. It will be hard to get them all to work perfectly – after all, we are talking about devices which are made on the nanoscale and which need to preserve quantum spookiness. Nothing can be allowed to measure or interfere with the system, either intentionally or not – otherwise it will destroy the double-life of right-handed and left-handed entanglement. Quantum gloves will become normal gloves, and quantum devices will become normal devices. So how can reliable devices be made? One can imagine the nightmare scenario of making a very large number of them, only to find at the end that no single device is good enough. It would be such a waste to throw them all away. It turns out that the same problem already arises with conventional computer chips – many are thrown away simply because they are too imperfect. So what can be done?

The answer is to form a crowd of them. Damien Challet and I have shown mathematically that by taking appropriate combinations of such defective devices, a much more accurate – indeed essentially perfect – device can be made. It turns out that the mathematical theory which we developed is intimately related to the complexity associated with combining numbers. In particular, it is related to the problem of taking combinations of numbers so that their sum is as close to zero as possible. But the basic idea can be understood quite simply in terms of a few clocks. Suppose you have a clock which is five minutes fast. If you want to produce an automatic readout of the time which is perfect, just connect it to a clock which is five minutes slow and arrange for an average of the readings to be displayed instead. The resulting time from this "crowd" of clocks is perfectly accurate. Indeed this is exactly what sailors used to do to deduce the correct time at sea. They simply took a collection of clocks on board and took the average or consensus of the times. This example considers everyday clocks – but there is no reason in principle why the same idea cannot be applied to nanoscale devices, even in the quantum regime. After all, everything has a readout eventually, and it is at this stage that taking a particular subset of devices whose errors effectively cancel, comes to the fore. Any remaining defective devices can then be recycled efficiently by repeating this process, thereby producing a further batch of devices which are themselves also quite accurate.

So let's see how this would work for a collection of such imperfect objects. Suppose we are clockmakers and we have produced six clocks whose times, relative to the exact time, are +5, +3, −8, −2, −1, +4. We need to compete with the big manufacturers whose clocks all have a guaranteed error of 1 or less. Should we throw away the clocks we have made and make new ones, thereby running the risk that we waste additional effort and money without producing anything more accurate? No, we just go ahead and form them into suitable "crowds". As shown in figure 11.2, we can form a crowd of the three clocks with errors +3, −2 and −1 in order to give a composite clock with a net error

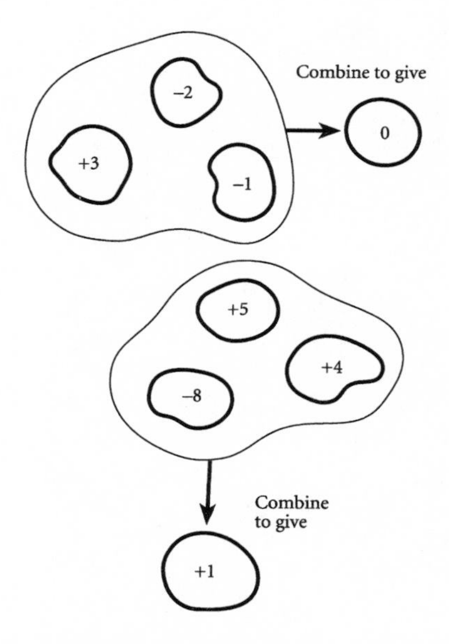

Figure 11.2 Making nothing out of something. By combining devices which are themselves inaccurate, far more accurate composite devices can be made. In the process, little or no waste is left.

of zero. In short, +3 − 2 − 1 = 0. We can then form a crowd of the three remaining clocks with errors +5, −8 and +4 to give a composite clock with net error +1, since +5 − 8 + 4 = +1. In addition to having used up all our otherwise useless clocks, we have actually produced two composite devices with near-perfect accuracy.

So it turns out that as a by-product of our interest in Complexity and hence crowds, we have actually come up with an efficient and ecological solution for dealing with wastage. In fact, it seems such a nice idea that one could imagine Nature already using it – or mankind being able to exploit it – to correct errors at the cellular level in human biology. In particular, might such a combination-of-errors approach be used to reduce the effects of cellular imperfections?

Chapter 12

To infinity and beyond

12.1 The inadequate infinity of physicists

We have seen in this book that the need to understand Complex Systems is motivated by a multitude of important practical applications, coupled with a very deep scientific significance. Examples of the collective phenomena which can emerge from real-world Complex Systems include traffic congestion, financial market crashes, wars, cancer and epidemics. Each of these phenomena represents a massive challenge to us as individuals and as a Society – from our daily commute home through to the performance of our pension funds, and from our daily health through to our life expectancy – since they can emerge spontaneously without any form of "invisible hand" or central controller. Such emergent phenomena are possible because the underlying system contains many interacting objects, and because there is some form of feedback in the system. That is why we need to understand what is going on in such Complex Systems both for scientific and for practical reasons.

Physics is used to dealing with large numbers of interacting objects. However, the answer to building a true *Theory of Complexity* is currently a bridge too far for Physics. Most physicists implicitly deal with closed systems, such as the Universe itself and systems which have reached some kind of steady state. Having said this, there is a branch of Physics which tries to avoid such assumptions – the field of non-equilibrium statistical mechanics. So is this the answer? Unfortunately, no. Or at least, not in its current form. The problem is that statistical mechanics in general tries to look at the limit of large numbers of objects – very large numbers, of the order of the number of atoms in a drop of liquid or a balloon full of air. This is fine for a drop of liquid or a balloon full of air since they really do have an amazingly large number of atoms in them. For example, a typical everyday volume would contain ten-to-the-power-twenty-something atoms, which is more than one hundred million, million, million. And this is much larger than the number of people on the planet.

It therefore seems unlikely that theories which need to assume such a large number of objects can properly represent everyday Complex Systems where the numbers involved are typically less than a thousand, or even a hundred. After all, in a financial market the number of people who actually have enough economic clout that they move the market when they trade is relatively few – and it is this number which should feature in any realistic model or theory of the market. So applying any theory which assumes that there are essentially an infinite number of such people sounds dodgy.

This practice by physicists of developing tools and theories which work when there are extraordinarily large numbers of objects, is certainly a very powerful one for liquids, gases and solids. As mentioned above, it works physically because these systems do indeed have lots of atoms in them – and it works mathematically because there is a tendency for large numbers of identical objects to behave like the average of them all. In addition, such theories typically assume some kind of "temperature" which implies that the system is in a steady-state. In other words, you have to wait a very long time to reach this state. But in the long run, humans are dead – so again this sounds dodgy. Therefore the notion of developing a theory which works for an infinite number of identical objects when they are in some kind of steady-state is fine – as long as that is the system that it is applied to. The hope that it can then be tweaked to apply to everyday Complex Systems which feature a finite number of non-identical objects, and which are not in a steady state, seems suspect – unfortunately.

12.2 The future is bright, the future is Complex

Things are, however, far from bleak for Complexity Science. Indeed there is a very bright future ahead in terms of the study of models and real-world systems which combine the two key manifestations of Complex Systems discussed throughout this book: multi-object or so-called multi-agent populations, and networks. For example, the possibility of using the decision-making within a multi-agent population in order to build and then manipulate a complex network, and then having that network feed back onto the decisions themselves, will be a very rich area for research since it is common to all the applications discussed in this book. In addition, there are many more applications which I have not mentioned waiting to be analyzed across the physical, biological and social sciences, and which cover a wide range of length-scales and time-scales: from the quantum scale right through to the structural properties of the Universe.

You can already see such research activity starting to happen, simply by typing *Complexity* or *Complex Systems* into Google. You will be flooded with lists of Workshops and Conferences focusing on a range of different topics and disciplines. Many of these aim to explore the competition or cooperation in groups or networks of decision-making agents, in order to see how this might underpin the dynamical evolution which is actually observed across the social, political, economic and biological spectrum. Examples include political instability, guerilla warfare, crime, foraging systems such as biological organisms (e.g. fungi), and animal herding phenomena. The common elements typically include the concept of an ecology of interacting agents, possibly foraging for some limited resource. In addition, these objects may be moving on some complex dynamical network, and indeed their own actions and evolution may themselves affect the network's structure and future evolution.

This increased focus on real-world Complexity is very good news since many professionals – from scientists through to medical doctors and policy-makers – already spend their days dealing with particular manifestations of Complexity yet may not even realize it. In other words, a better appreciation of the ideas behind

Complexity, as discussed in this book, could help us all reap significant practical benefits.

In terms of academic advancement, there seems to be an increasing proportion of the research community which is taking on board the idea that there will be few major advances in human medicine, sociology or economics without a better understanding and appreciation of Complexity Science. For example, large areas of biological phenomena are beginning to be lumped together under the more general Complexity-related title of "Systems Biology". More medically-orientated projects seem to be referred to as "Systems Medicine", "Nanobiomedicine" or some other similar hybrid. Even in genetics there is plenty of room for using ideas from Complexity. In particular, the vast amounts of DNA code that have been measured are supposed to read like a book, with each gene representing a phrase. Yet we all know that the true meaning of any book lies in the overall combination of such phrases, and in particular their interactions. Hence the true meaning of our genetic code is likely to lie in the collective behavior of this crowd of genes – and this collective behavior will arise, as with any Complex System, through interactions and feedback. Without it, the DNA book may be readable – but nobody will understand what it means.

So as we have seen throughout this book, Complexity isn't just important for understanding traffic jams, financial market crashes or cancer growth. It also lies at the heart of the Universe itself. It is therefore "Big Science". However unlike all previous "Big Science", it also has tremendous everyday importance – from our own personal health, wealth and lifestyle, through to the security and prosperity of our Society as a whole. *Complexity is indeed the science of all sciences.*

Appendix

Further Information

This Appendix provides further information for anyone interested in pursuing the issues, topics and research discussed in this book. In section B, I provide details of how to access the research papers themselves. However, by their very nature research papers tend to be written in a very concise style and often contain very specific terminology – so it can be hard work jumping straight into reading them. Therefore, section A provides a stepping stone in the form of general webpages on Complex Systems, Complexity and centers of study around the world, together with a list of popular science books on these topics.

A. Complexity, Complex Systems and centers of study

By far the easiest way of obtaining up-to-date information is to type "Complexity" or "Complex Systems" into any Internet search engine, e.g. Google. Below are some Internet sites which can be used as a starting point for exploring Complexity. The list is not exhaustive – nor can I endorse individual sites or guarantee whether they are currently accurate or up-to-date. However, they do supplement the discussions in this book and provide a broad picture of how this new field is evolving.

Tutorials and software

See the following short tutorial-style accounts on Wikipedia:
Complexity: *http://en.wikipedia.org/wiki/Complexity*
Chaos Theory: *http://en.wikipedia.org/wiki/Chaos_theory*
Randomness: *http://en.wikipedia.org/wiki/Randomness*
Quantum Physics: *http://en.wikipedia.org/wiki/Quantum_ mechanics*
Entanglement: *http://en.wikipedia.org/wiki/Quantum_entanglement*
Superposition: *http://en.wikipedia.org/wiki/Quantum_superposition*
Cancer: *http://en.wikipedia.org/wiki/Cancer*
Angiogenesis: *http://en.wikipedia.org/wiki/Angiogenesis*

In addition, the following sites are worth visiting:

http://www.calresco.org/tutorial.htm – A source of tutorials on chaos, fractals and more general Complex Systems topics

http://cognitrn.psych.indiana.edu/rgoldsto/complex/ – Software for demonstrating Complex Systems from researchers at Indiana University, U.S.A.

http://necsi.org/guide/index.html – A general guide from the New England Complex Systems Institute, with additional online resources at *http://necsi.org/education/onlineproj.html* and simulations at *http://necsi.org/visual/*

Updates of recent developments in Complexity Science

http://www.comdig.org/ – A digest which provides an overview of new developments.

See also the list of resources at
http://www.comdig.org/resources.php

http://www.complexity-society.com/ – The Complexity Society which looks at the application of Complexity Science to human affairs.

General books

For those interested primarily in financial applications, I suggest looking at:

N.F. Johnson, P. Jefferies and P.M. Hui, *Financial Market Complexity* (OUP, Oxford, 2003) ISBN: 0198526652

For broader discussions on topics such as non-linearity, chaos, power-laws and collective behavior, I recommend looking at the following popular science books:

Philip Ball, *Critical Mass: How One Thing Leads to Another* (Arrow Books, 2005) ISBN: 0099457865

Mark Buchanan, *Nexus: Small Worlds and the Groundbreaking Theory of Networks* (Norton Books, 2003) ISBN: 0393324427

Mark Buchanan, *Ubiquity: The New Science That Is Changing the World* (Phoenix Press, 2001) ISBN: 0753812975

Ricard Sole, Brian Goodwin, *Signs of Life: How Complexity Pervades Biology* (Basic Books, 2002) ISBN: 0465019285

Albert-Laszlo Barabasi, *Linked: How Everything Is Connected to Everything Else and What It Means for Business, Science, and Everyday Life* (Plume Books, 2003) ISBN: 0452284392

Steven Johnson, *Emergence: The Connected Lives of Ants, Brains, Cities and Software* (Penguin Books, 2002) ISBN: 0140287752

Roger Lewin, *Complexity: Life at the Edge of Chaos* (Phoenix Press, 2001) ISBN: 0753812703

Steven Strogatz, *Sync: The Emerging Science of Spontaneous Order* (Penguin Books, 2004) ISBN: 014100763X

Duncan J. Watts, *Small Worlds: The Dynamics of Networks Between Order and Randomness* (Princeton University Press, 2004) ISBN: 0691117047

John Scott, *Social Network Analysis: A Handbook* (Sage Publications, 2000) ISBN: 0761963391

Duncan J. Watts, *Six Degrees: The New Science of Networks* (Vintage Books, 2004) ISBN: 0099444968

Per Bak, *How Nature Works: The Science of Self-Organized Criticality* (Springer-Verlag, 1996) ISBN: 0387947914

John Gribbin, *Deep Simplicity: Chaos Complexity and the Emergence of Life* (Penguin Books, 2005) ISBN: 0141007222

John Gribbin, *Schrodinger's Kittens and the Search for Reality: The Quantum Mysteries Solved* (Phoenix Books, 1996) ISBN: 1857994027

Eric Bonabeau, Marco Dorigo, Guy Theraulaz, *Swarm Intelligence: From Natural to Artificial Systems* (Oxford University Press, 1999) ISBN: 0195131592

Centers for Complexity: real and virtual

http://sbs-xnet.sbs.ox.ac.uk/complexity/complexity_home.asp – Complex Systems homepage for the University of Oxford, U.K.

http://www.santafe.edu/ – The Santa Fe Institute in New Mexico, U.S.A., which is a multidisciplinary research center

http://ti.arc.nasa.gov/ – The Intelligent Systems Division at NASA

http://www.cscs.umich.edu/ – The University of Michigan's Center for the Study of Complex Systems

http://www.ccs.fau.edu/ – Florida Atlantic University's Center for Complex Systems and Brain Sciences

B. Downloadable research papers

The research papers listed below contain the background details for the discussions in this book, as well as reports on related areas of research. The references which are cited at the end of each of these papers provide further information about the publications by other research groups in the relevant area. Further details of my own group's research collaborations, and an update of the projects described in this book, can be found on the following websites:

http://sbs-xnet.sbs.ox.ac.uk/complexity/complexity_home.asp
http://users.physics.ox.ac.uk/cmphys/cmt/people.htm

Each paper listed below can be freely obtained online by visiting **http://xxx.lanl.gov**. In particular:

To access paper **physics/0604121**, for example, go to the following website:
http://xxx.lanl.gov/abs/physics and type in the paper number **0604121**

To access paper **quant-ph/0509022**, for example, go to the following website:
http://xxx.lanl.gov/abs/quant-ph and type in the paper number **0509022**

To access paper **cond-mat/0506011**, for example, go to the following website:
http://xxx.lanl.gov/abs/cond-mat and type in the paper number **0506011**

The same procedure applies for any paper number. Nearly all the papers listed below are already published in peer-reviewed, international science journals. However, I won't give the reference to these journals since access to them is not generally free. Additional research papers written by the wider Complexity community can be freely downloaded from the following two websites:

http://xxx.lanl.gov
http://www.unifr.ch/econophysics

1. physics/0604121
   Title: *Multi-Agent Complex Systems and Many-Body Physics*
   Authors: N.F. Johnson, David M.D. Smith, Pak Ming Hui
2. physics/0605035
   Title: *Universal patterns underlying ongoing wars and terrorism*
   Authors: N.F. Johnson, Mike Spagat, Jorge A. Restrepo, Oscar Becerra, Juan Camilo Bohorquez, Nicolas Suarez, Elvira Maria Restrepo, Roberto Zarama
3. physics/0604142
   Title: *Pair Formation within Multi-Agent Populations*
   Authors: David M.D. Smith, N.F. Johnson
4. physics/0605065
   Title: *Predictability, Risk and Online Management in a Complex System of Adaptive Agents*
   Authors: David M.D. Smith, N.F. Johnson
5. physics/0604183
   Title: *Interplay between function and structure in complex networks*
   Authors: Timothy C. Jarrett, Douglas J. Ashton, Mark Fricker, N.F. Johnson
6. physics/0604148
   Title: *Effects of decision-making on the transport costs across complex networks*
   Authors: Sean Gourley, N.F. Johnson
7. cond-mat/0604623
   Title: *Optically controlled spin-glasses in multi-qubit cavity systems*
   Authors: Timothy C. Jarrett, Chiu Fan Lee, N.F. Johnson

8. quant-ph/0509022
   Title: *Renormalization scheme for a multi-qubit-network*
   Authors: Alexandra Olaya-Castro, Chiu Fan Lee, N.F. Johnson
9. physics/0508228
   Title: *Abrupt structural transitions involving functionally optimal networks*
   Authors: Timothy C. Jarrett, Douglas J. Ashton, Mark Fricker, N.F. Johnson
10. quant-ph/0507164
    Title: *Reply to Brankov et al.'s "Comment on equivalence between quantum phase transition phenomena in radiation-matter and magnetic systems"*
    Authors: J. Reslen, L. Quiroga, N.F. Johnson
11. physics/0506213
    Title: *From old wars to new wars and global terrorism*
    Authors: N.F. Johnson, M. Spagat, J. Restrepo, J. Bohorquez, N. Suarez, E. Restrepo, R. Zarama
12. physics/0506134
    Title: *Using Artificial Market Models to Forecast Financial Time-Series*
    Authors: Nachi Gupta, Raphael Hauser, N.F. Johnson
13. cond-mat/0506011
    Title: *A non-Markovian optical signature for detecting entanglement in coupled excitonic qubits*
    Authors: F. J. Rodriguez, L. Quiroga, N.F. Johnson
14. cond-mat/0505581
    Title: *Decision Making, Strategy dynamics, and Crowd Formation in Agent-based models of Competing Populations*
    Authors: K.P. Chan, Pak Ming Hui, N.F. Johnson
15. cond-mat/0505575
    Title: *Transitions in collective response in multi-agent models of competing populations driven by resource level*
    Authors: Sonic H. Y. Chan, T. S. Lo, P. M. Hui, N.F. Johnson
16. physics/0505071
    Title: *How does Europe Make Its Mind Up? Connections, cliques, and compatibility between countries in the Eurovision Song Contest*

Authors: Daniel Fenn, Omer Suleman, Janet Efstathiou, N.F. Johnson

17. physics/0503031
Title: *Competitive Advantage for Multiple-Memory Strategies in an Artificial Market*
Authors: Kurt E. Mitman, Sehyo Charley Choe, N.F. Johnson
18. quant-ph/0503015
Title: *Exploring super-radiant phase transitions via coherent control of a multi-qubit–cavity system*
Authors: Timothy C. Jarrett, Chiu Fan Lee, N.F. Johnson
19. physics/0503014
Title: *What shakes the FX tree? Understanding currency dominance, dependence and dynamics*
Authors: N.F. Johnson, Mark McDonald, Omer Suleman, Stacy Williams, Sam Howison
20. cond-mat/0501186
Title: *Many-Body Theory for Multi-Agent Complex Systems*
Authors: N.F. Johnson, David M.D. Smith, Pak Ming Hui
21. cond-mat/0412411
Title: *Detecting a Currency's Dominance or Dependence using Foreign Exchange Network Trees*
Authors: Mark McDonald, Omer Suleman, Stacy Williams, Sam Howison, N.F. Johnson
22. quant-ph/0412069
Title: *Optically controlled spin-glasses generated using multi-qubit cavity systems*
Authors: Chiu Fan Lee, N.F. Johnson
23. cond-mat/0409140
Title: *Theory of enhanced performance emerging in a sparsely-connected competitive population*
Authors: T.S. Lo, K.P Chan, P.M. Hui, N.F. Johnson
24. quant-ph/0409104
Title: *A robust one-step catalytic machine for high fidelity anticloning and W-state generation in a multi-qubit system*
Authors: Alexandra Olaya-Castro, N.F. Johnson, Luis Quiroga

25. cond-mat/0409059
Title: *Effect of congestion costs on shortest paths through complex networks*
Authors: Douglas J. Ashton, Timothy C. Jarrett, N.F. Johnson

26. cond-mat/0409036
Title: *Evolution Management in a Complex Adaptive System: Engineering the Future*
Authors: David M.D. Smith, N.F. Johnson

27. cond-mat/0408557
Title: *Plateaux formation, abrupt transitions, and fractional states in a competitive population with limited resources*
Authors: H. Y. Chan, T. S. Lo, P. M. Hui, N.F. Johnson

28. cond-mat/0406674
Title: *Direct equivalence between quantum phase transition phenomena in radiation-matter and magnetic systems: scaling of entanglement*
Authors: José Reslen, Luis Quiroga, N.F. Johnson

29. cond-mat/0406391
Title: *Theory of Networked Minority Games based on Strategy Pattern Dynamics*
Authors: T. S. Lo, H. Y. Chan, P. M. Hui, N.F. Johnson

30. quant-ph/0406133
Title: *Quantum Information Processing in Nanostructures*
Authors: Alexandra Olaya-Castro, N.F. Johnson

31. cond-mat/0405037
Title: *Error-driven Global Transition in a Competitive Population on a Network*
Authors: Sehyo Charley Choe, N.F. Johnson, Pak Ming Hui

32. quant-ph/0404163
Title: *Efficient quantum computation within a disordered Heisenberg spin-chain*
Authors: Chiu Fan Lee, N.F. Johnson

33. quant-ph/0403185
Title: *First-order super-radiant phase transitions in a multi-qubit–cavity system*
Authors: Chiu Fan Lee, N.F. Johnson

34. cond-mat/0403158
Title: *Theory of Collective Dynamics in Multi-Agent Complex Systems*
Authors: N.F. Johnson, Sehyo C. Choe, Sean Gourley, Timothy Jarrett, Pak Ming Hui
35. cond-mat/0401527
Title: *Dynamical interplay between local connectivity and global competition in a networked population*
Authors: S. Gourley, S.C. Choe, P.M. Hui, N.F. Johnson
36. cond-mat/0312556
Title: *Memory and self-induced shocks in an evolutionary population competing for limited resources*
Authors: Roland Kay, N.F. Johnson
37. cond-mat/0312321
Title: *Enhanced Winning in a Competing Population by Random Participation*
Authors: K.F. Yip, T.S. Lo, P.M. Hui, N.F. Johnson
38. quant-ph/0311009
Title: *Quantum random walks with history dependence*
Authors: Adrian P. Flitney, Derek Abbott, N.F. Johnson
39. cond-mat/0306516
Title: *Crowd-Anticrowd Theory of Collective Dynamics in Competitive, Multi-Agent Populations and Networks*
Authors: N.F. Johnson, Pak Ming Hui
40. cond-mat/0212505
Title: *Interacting many-body systems as non-cooperative games*
Authors: Chiu Fan Lee, N.F. Johnson
41. cond-mat/0212088
Title: *Crowd-Anticrowd Theory of Multi-Agent Minority Games*
Authors: Michael L. Hart, N.F. Johnson
42. quant-ph/0210192
Title: *Non-Cooperative Quantum Game Theory*
Authors: Chiu Fan Lee, N.F. Johnson
43. quant-ph/0210185
Title: *Quantum coherence, correlated noise and Parrondo games*
Authors: Chiu Fan Lee, N.F. Johnson, Ferney Rodriguez, Luis Quiroga

44. cond-mat/0210132
Title: *Managing catastrophic changes in a collective*
Authors: David Lamper, Paul Jefferies, Michael Hart, N.F. Johnson

45. cond-mat/0207588
Title: *An Investigation of Crash Avoidance in a Complex System*
Authors: Michael L. Hart, David Lamper, N.F. Johnson

46. cond-mat/0207523
Title: *Designing agent-based market models*
Authors: Paul Jefferies, N.F. Johnson

47. cond-mat/0207386
Title: *Winning combinations of history-dependent games*
Authors: Roland J. Kay, N.F. Johnson

48. quant-ph/0207139
Title: *Game-theoretic discussion of quantum state estimation and cloning*
Authors: Chiu Fan Lee, N.F. Johnson

49. quant-ph/0207080
Title: *Exploiting Randomness in Quantum Information Processing*
Authors: Chiu Fan Lee, N.F. Johnson

50. quant-ph/0207012
Title: *Quantum Game Theory*
Authors: Chiu Fan Lee, N.F. Johnson

51. cond-mat/0206228
Title: *Crash Avoidance in a Complex System*
Authors: Michael L. Hart, David Lamper, N.F. Johnson

52. quant-ph/0203043
Title: *Parrondo Games and Quantum Algorithms*
Authors: Chiu Fan Lee, N.F. Johnson

53. cond-mat/0203028
Title: *Optimal combinations of imperfect objects*
Authors: D. Challet, N.F. Johnson

54. cond-mat/0201540
Title: *Anatomy of extreme events in a complex adaptive system*
Authors: Paul Jefferies, David Lamper, N.F. Johnson

55. cond-mat/0112501
    Title: *Herd Formation and Information Transmission in a Population: Non-universal behavior*
    Authors: Dafang Zheng, P. M. Hui, K. F. Yip, N.F. Johnson
56. cond-mat/0105474
    Title: *Non-universal scaling in a model of information transmission and herd behavior*
    Authors: Dafang Zheng, P. M. Hui, N.F. Johnson
57. cond-mat/0105303
    Title: *Application of multi-agent games to the prediction of financial time-series*
    Authors: N.F. Johnson, D. Lamper, P. Jefferies, M. L. Hart, S. Howison
58. cond-mat/0105258
    Title: *Predictability of large future changes in a competitive evolving population*
    Authors: D. Lamper, S. Howison, N.F. Johnson
59. quant-ph/0105029
    Title: *Decoherence of quantum registers*
    Authors: John H. Reina, Luis Quiroga, N.F. Johnson
60. cond-mat/0103259
    Title: *Deterministic Dynamics in the Minority Game*
    Authors: P. Jefferies, M.L. Hart, N.F. Johnson
61. cond-mat/0102384
    Title: *Dynamics of the Time Horizon Minority Game*
    Authors: Michael L. Hart, Paul Jefferies, N.F. Johnson
62. quant-ph/0102008
    Title: *Evolutionary quantum game*
    Authors: Roland Kay, N.F. Johnson, Simon C. Benjamin
63. quant-ph/0009050
    Title: *Playing a quantum game with a corrupted source*
    Author: N.F. Johnson
64. quant-ph/0009035
    Title: *Quantum information processing in semiconductor nanostructures*
    Authors: John H. Reina, Luis Quiroga, N.F. Johnson

65. cond-mat/0008387
Title: *From market games to real-world markets*
Authors: P. Jefferies, M.L. Hart, P.M. Hui, N.F. Johnson

66. cond-mat/0008385
Title: *Crowd-Anticrowd Theory of Multi-Agent Market Games*
Authors: M. Hart, P. Jefferies, P.M. Hui, N.F. Johnson

67. cond-mat/0006141
Title: *Stochastic strategies in the Minority Game*
Authors: M. Hart, P. Jefferies, N.F. Johnson, P.M. Hui

68. cond-mat/0006122
Title: *Evolutionary minority game with heterogeneous strategy distribution*
Authors: T.S. Lo, S.W. Lim, P.M. Hui, N.F. Johnson

69. cond-mat/0005152
Title: *Crowd-anticrowd theory of the Minority Game*
Authors: M. Hart, P. Jefferies, N.F. Johnson, P.M. Hui

70. cond-mat/0005043
Title: *Mixed population Minority Game with generalized strategies*
Authors: P. Jefferies, M. Hart, N.F. Johnson, P.M. Hui

71. cond-mat/0004063
Title: *Generalized strategies in the Minority Game*
Authors: M. Hart, P. Jefferies, N.F. Johnson, P.M. Hui

72. cond-mat/0003486
Title: *Crowd-anticrowd model of the Minority Game*
Authors: M. Hart, P. Jefferies, N.F. Johnson, P.M. Hui

73. cond-mat/0003379
Title: *Theory of the evolutionary minority game*
Authors: T.S. Lo, P.M. Hui, N.F. Johnson

74. cond-mat/0003309
Title: *Segregation in a competing and evolving population*
Authors: P.M. Hui, T.S. Lo, N.F. Johnson

75. cond-mat/9910072
Title: *Trader dynamics in a model market*
Authors: N.F. Johnson, Michael Hart, Pak Ming Hui, Dafang Zheng

76. cond-mat/9909139
Title: *Decoherence effects on the generation of exciton entangled states in coupled quantum dots*
Authors: F.J. Rodriguez, L. Quiroga, N.F. Johnson
77. cond-mat/9906034
Title: *Quantum Teleportation in a Solid State System*
Authors: John H. Reina, N.F. Johnson
78. cond-mat/9905039
Title: *Evolutionary freezing in a competitive population*
Authors: N.F. Johnson, D.J.T. Leonard, P.M. Hui, T.S. Lo
79. cond-mat/9903228
Title: *Minority game with arbitrary cutoffs*
Authors: N.F. Johnson, P.M. Hui, Dafang Zheng, C.W. Tai
80. cond-mat/9903164
Title: *Enhanced winnings in a mixed-ability population playing a minority game*
Authors: N.F. Johnson, P.M. Hui, D. Zheng, M. Hart
81. cond-mat/9901201
Title: *Entangled Bell and GHZ states of excitons in coupled quantum dots*
Authors: Luis Quiroga, N.F. Johnson
82. cond-mat/9811227
Title: *Crowd effects and volatility in a competitive market*
Authors: N.F. Johnson, Michael Hart, Pak Ming Hui
83. cond-mat/9810142
Title: *Self-Organized Segregation within an Evolving Population*
Authors: N.F. Johnson, Pak Ming Hui, Rob Jonson, Ting Shek Lo
84. cond-mat/9808243
Title: *Cellular Structures for Computation in the Quantum Regime*
Authors: S. C. Benjamin, N.F. Johnson
85. cond-mat/9802177
Title: *Volatility and Agent Adaptability in a Self-Organizing Market*
Authors: N.F. Johnson, S. Jarvis, R. Jonson, P. Cheung, Y.R. Kwong, P.M. Hui

Index

Note: Page references in *italics* indicate illustrations.

Index

Index

Index

Index

Index